MANAGING FOR WORLD-CLASS QUALITY

QUALITY AND RELIABILITY

A Series Edited by

1. Designing for Minimal Maintenance Expense: The Practical Application of Reliability and Maintainability, *Marvin A. Moss*
2. Quality Control for Profit, Second Edition, Revised and Expanded, *Ronald H. Lester, Norbert L. Enrick, and Harry E. Mottley, Jr.*
3. QCPAC: Statistical Quality Control on the IBM PC, *Steven M. Zimmerman and Leo M. Conrad*
4. Quality by Experimental Design, *Thomas B. Barker*
5. Applications of Quality Control in the Service Industry, *A. C. Rosander*
6. Integrated Product Testing and Evaluating: A Systems Approach to Improve Reliability and Quality, Revised Edition, *Harold L. Gilmore and Herbert C. Schwartz*

7. Quality Management Handbook, *edited by Loren Walsh, Ralph Wurster, and Raymond J. Kimber*
8. Statistical Process Control: A Guide for Implementation, *Roger W. Berger and Thomas Hart*
9. Quality Circles: Selected Readings, *edited by Roger W. Berger and David L. Shores*
10. Quality and Productivity for Bankers and Financial Managers, *William J. Latzko*
11. Poor-Quality Cost, *H. James Harrington*
12. Human Resources Management, *edited by Jill P. Kern, John J. Riley, and Louis N. Jones*
13. The Good and the Bad News About Quality, *Edward M. Schrock and Henry L. Lefevre*
14. Engineering Design for Producibility and Reliability, *John W. Priest*
15. Statistical Process Control in Automated Manufacturing, *J. Bert Keats and Norma Faris Hubele*
16. Automated Inspection and Quality Assurance, *Stanley L. Robinson and Richard K. Miller*
17. Defect Prevention: Use of Simple Statistical Tools, *Victor E. Kane*
18. Defect Prevention: Use of Simple Statistical Tools, Solutions Manual, *Victor E. Kane*
19. Purchasing and Quality, *Max McRobb*
20. Specification Writing and Management, *Max McRobb*
21. Quality Function Deployment: A Practitioner's Approach, *James L. Bossert*
22. The Quality Promise, *Lester Jay Wollschlaeger*
23. Statistical Process Control in Manufacturing, *edited by J. Bert Keats and Douglas C. Montgomery*
24. Total Manufacturing Assurance, *Douglas C. Brauer and John Cesarone*
25. Deming's 14 Points Applied to Services, *A. C. Rosander*
26. Evaluation and Control of Measurements, *John Mandel*
27. Achieving Excellence in Business: A Practical Guide to the Total Quality Transformation Process, *Kenneth E. Ebel*
28. Statistical Methods for the Process Industries, *William H. McNeese and Robert A. Klein*

29. Quality Engineering Handbook, *edited by Thomas Pyzdek and Roger W. Berger*
30. Managing for World-Class Quality: A Primer for Executives and Managers, *Edwin S. Shecter*
31. A Leader's Journey to Quality, *Dana M. Cound*

Additional volumes in preparation

MANAGING FOR WORLD-CLASS QUALITY

A Primer for Executives and Managers

EDWIN S. SHECTER

Quality Resources Company
Lawrenceville, New Jersey

Marcel Dekker, Inc.
New York • Basel • Hong Kong

ASQC Quality Press
Milwaukee

Library of Congress Cataloging-in-Publication Data

Shecter, Edwin.
Managing for world-class quality : a primer for executives and managers / Shecter, Edwin S.
p. cm. — (Quality and reliability ; 30)
Includes bibliographical references and index.
ISBN 0-8247-7712-3 (alk. paper)
1. Total quality management. 2. Quality control. 3. Reliability (Engineering) I. Title. II. Series.
HD62.15.S54 1991
658.5′62—dc20 91-22455
CIP

This book is printed on acid-free paper.

Marcel Dekker, Inc.
270 Madison Avenue, New York, New York 10016

ASQC Quality Press
310 West Wisconsin Avenue
Milwaukee, Wisconsin 53203

Current printing (last digit):
10 9 8 7 6 5 4 3 2 1

PRINTED IN THE UNITED STATES OF AMERICA

I dedicate this book to my wife

CELIA

without whose continuing support and encouragement

this book would not have been possible, and to

DEBBIE *and* **LORI**

for their unflagging love and affection.

About the Series

The genesis of modern methods of quality and reliability will be found in a simple memo dated May 16, 1924, in which Walter A. Shewhart proposed the control chart for the analysis of inspection data. This led to a broadening of the concept of inspection from emphasis on detection and correction of defective material to control of quality through analysis and prevention of quality problems. Subsequent concern for product performance in the hands of the user stimulated development of the systems and techniques of reliability. Emphasis on the consumer as the ultimate judge of quality serves as the catalyst to bring about the integration of the methodology of quality with that of reliability. Thus, the innovations that came out of the control chart spawned a philosophy of control of quality and reliability that has come to include not only the methodology of the statistical sciences and engineering, but also the use of appropriate management methods together with various motivational procedures in a concerted effort dedicated to quality improvement.

This series is intended to provide a vehicle to foster interaction of the elements of the modern approach to quality, including statistical applications, quality and reliability engineering, management, and motivational aspects. It is a forum in which the subject matter of these various areas can be brought together to allow for effective integration of appropriate techniques. This will promote the true benefit of each, which can be achieved only through their interaction. In this sense, the whole of quality and reliability is greater than the sum of its parts, as each element augments the others.

The contributors to this series have been encouraged to discuss fundamental concepts as well as methodology, technology, and procedures at the leading edge of the discipline. Thus, new concepts are placed in proper perspective in these evolving disciplines. The series is intended for those in manufacturing, engineering, and marketing and management, as well as the consuming public, all of whom have an interest and stake in the improvement and maintenance of quality and reliability in the products and services that are the lifeblood of the economic system.

The modern approach to quality and reliability concerns excellence: excellence when the product is designed, excellence when the product is made, excellence as the product is used, and excellence throughout its lifetime. But excellence does not result without effort, and products and services of superior quality and reliability require an appropriate combination of statistical, engineering, management, and motivational effort. This effort can be directed for maximum benefit only in light of timely knowledge of approaches and methods that have been developed and are available in these areas of expertise. Within the volumes of this series, the reader will find the means to create, control, correct, and improve quality and reliability in ways that are cost effective, that enhance productivity, and that create a motivational atmosphere that is harmonious and constructive. It is dedicated to that end and to the readers whose study of quality and reliability will lead to greater understanding of their products, their processes, their workplaces, and themselves.

Edward G. Schilling

Preface

This book has been written to share the lessons learned during more than thirty years in industry working in the quality control field. It is intended to provide insight into problems and suggest solutions to these problems.

It is my contention that proper attention to the quality function in American industry can result in enormous economic benefits to companies using this philosophy and methodology, and can help prevent economic recessions in the United States.

The emphasis placed on the quality function in Japan has enabled the Japanese to change their product quality image and resulted in major market penetration in many industries, thereby forcing cutbacks in U.S. employment. The methods that they used for accomplishing this were taught to them by Americans who were not understood by top management in the United States. Perhaps this is understandable, since the Japanese top manager is more of a generalist, while the American top manager is more of a specialist. As such, it is necessary to communicate with American management in the language of the specialist, and this can be difficult.

This book attempts to present information in a way that should help the manager understand the concept of quality. It uses examples—drawn from experience—of a comprehensive quality program that can be employed by all companies to accomplish their goals. This is not a book on statistics, although some statistical concepts are used. These are presented primarily to show how their use can improve product quality.

The concepts offered are intended to give the reader insight into what to look for in a quality program. Better product quality must result in lower costs or the quality program fails. Too often, in the past, quality professionals ignored the cost impact of poor quality and dealt only in statistical measures such as percent of nonconformance or defects per unit. Thus, the true effect of these data on company performance was not apparent.

The fact is that good quality and low cost go hand in hand. Good product quality begets market share and customer loyalty. And customer loyalty spills over into all products made by a company. A side benefit, but a significant factor, is that worker attitudes toward a company turning out a high-quality product (one that satisfies the user's needs and expectations) are more positive and supportive. A good quality program improves the productivity, profitability, and stability of a company. You don't believe it? Read on and see if there aren't some tips that can help you and your organization.

Edwin S. Shecter

Contents

MANAGING FOR WORLD-CLASS QUALITY

1 · Introduction to Quality

This book is being written to share experiences I have had in making companies more competitive and profitable. I have read many books and countless articles on quality, productivity, and profitability. Yet there seem to be activities with which I was directly involved that applied a thought process coupled with some quality control techniques I have not seen widely discussed. One company where these methods were applied was so successful in developing products that it drove a long-established competitor into seeking a buyer to stay competitive.

The underlying philosophy, which I heard expressed many years ago, is that it is never sufficient to run a business to produce goods or services. One must produce information at the same time and use this information to improve the business. There is always concern on the part of managers who first hear this philosophy that a great deal of capital is required to achieve this operational practice. My experience has been that the opposite is true. Yes, there are times when an investment is necessary. Often, however, all that is necessary is to change an operating practice or some processing parameters to achieve major gains; those resulting in cost reduction, product or service improvement, higher productivity, and the attendant profit increase and market gains.

I was fortunate early in my career to be exposed to Ellis R. Ott, under whose tutelage creative ideas not only flourished but were cultivated. I was equally fortunate to work for Gus Mundel, who helped structure company philosophy. He preached that we could achieve great things using statistical quality control. He believed that this achievement would provide great economic gains for the company. This was accomplished, but it was unfortunate that top management did not encourage this performance to allow the company to meet its full potential.

Let me cite some examples of performance improvements and illustrate how they were accomplished. Since I was in the electronics business, most of these experiences are related to that business, but along the way I have experienced other examples that are worth recounting. In the following chapters, I present what I believe are the underlying fundamentals. The essential methodology that can be applied enables us to compete successfully in the world marketplace.

First, let me define what is meant by *world-class quality:* It is the production of goods or services that not only are the quality and price leaders in one country but which are capable of capturing and retaining market share elsewhere in the world on the basis of performance superior to the competition, as well as providing trouble-free service to customers up to and beyond their expectations. The real challenge is to accomplish this without diminishing the quality of life. This means, for example, that higher labor costs in the United States must be offset by significantly greater productivity. The approaches described in this book cannot, of course, overcome restrictive trade practices in foreign countries, nor can they accommodate political factors or monetary policies that preclude export. However, the provider of goods or services of world-class quality should have a major advantage over foreign competitors in the domestic marketplace and a significant edge over other domestic sources that do not use these methods. There are many other aspects of business in marketing, finance, research, development, and engineering that are needed to make a business successful. By proper application of the practices described in this book to these and other operational areas, dramatic benefits should be realized.

We in the United States simply cannot afford to play catch-up. We must play leap frog to compete successfully in the global marketplace. Furthermore, I do not believe that reducing the value of the dollar will result in correcting the trade imbalance. All that tactic will succeed in accomplishing is to make foreign trading partners more efficient and the United States less competitive. To achieve long-term results we must continually improve all aspects of our business, not the least of which is product or service quality.

One more term needing definition is *process*. A process is the organization of people, equipment, computer programs, material, procedures or systems, and facilities needed or intended to produce a specific end result, either a product or a service. A process can range from a single operation involving one machine, one operator, or no operators, and a series of parts or materials whose condition is changed to a particular end result, to one that involves many individual operations, starting with raw materials and ending with a final product.

If a process can be made to perform within specification limits, the process is said to be *capable;* alternatively, we say that the process capability is well within specification limits. When this situation exists, inspection and test can be reduced or eliminated and costs will be minimized. One struggle that is never-ending is to reduce process variability so that defects will not be generated.

A process can consist of a business system or a part of a business system. More comprehensive processes include manufacturing a fairly complex product such as a television receiver, refrigerator, automobile, or even a satellite. A process in a business system might involve processing a purchase order. A broader aspect determines the ordering cycle, including the purchasing process, identifying the needs of the company to initiate a purchase requisition for the correct quantity, type of material, specification, and other factors leading to receipt and actual use of the item procured. One primary concern of a business should be its operating cycle time: the period from receipt of an order to delivery (and installation) of the item into the customer's hands. Still another type of business cycle is getting a product or service from concept to the market. These cycles must be reduced in time to be successful in the global marketplace.

These processes all have in common the fact that they involve a number of variables, including people, machines, equipment or materials, and a system for combining these into useful output. We discuss concepts designed to improve the productivity, yield, conformance, or efficiency of the process to assure excellent output at low cost.

Quality Control at Work

Before proceeding, it might be helpful to examine some case histories.

Case History 1: Cutting Plates for Nickel-Cadmium Battery Manufacture

In the manufacture of nickel–cadmium batteries, each battery consists of a series of cells containing plates. A nickel–cadmium battery plate is made by sintering carbonyl nickel onto a nickel mesh very much like a fly screen. The plate is about 30 thousandths of an inch thick and can range in size from that of a playing card to typewriter paper, depending on the battery capacity or number of ampere-hours necessary for the application. Plates are processed in larger sizes called plaques, which undergo chemical treatment and then get cut to final size using a shear press. Positive and negative plates are alternated and separated by insulation to form the positive and negative terminals of the cell. It is important to prevent the positive plates from contacting any negative plates, so the edges and surface of the plates must be smooth. To contain the wires in the plate, the sintered plaque is compressed, or coined, in a pattern that provides the outline of the plate (see Fig. 1.1). Since

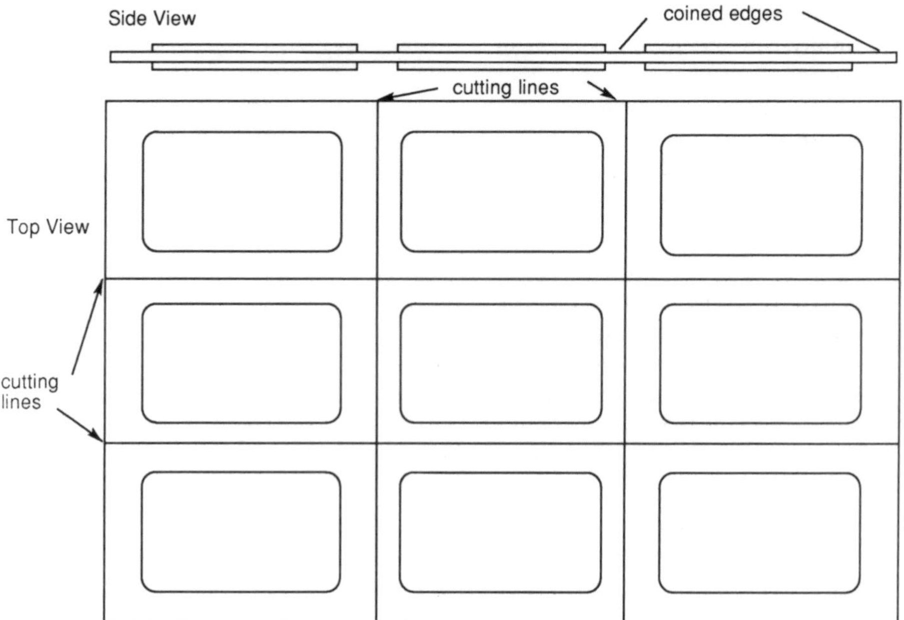

Figure 1.1 Nickel–cadmium battery plaque.

the cutting process requires a great deal of precision, a stop was provided on the shear to enable the cutting operator to guide the plaque precisely under the shear blade. Unfortunately, variations in the process can cause loose wires from the screen to break free. Then plates crack and cause rough surfaces and a host of other problems. The result can be a reject rate as high as about 50%.

In our example, originally, there was one operator cutting plates. To keep production moving, it became necessary to sort all the plates using one inspector and one or two rework operators to repair the very expensive damaged plates. As production increased, an extra cutting operator was added and an extra inspector was needed, together with four or five more rework people. Control charts were maintained to record the fraction defective (see Chapter 7). The process of cutting plaques into plates was under statistical control but was operating at the wrong level (i.e., 50% defective). Nonetheless, this operation continued to perform in this fashion for years and most people thought this was the best the process could do. Plant management resisted efforts to improve the process, stating that it was a waste of time and money to try to improve those conditions.

The quality manager persisted in trying to improve the situation and finally convinced the production manager and manufacturing engineering manager to form a team to try to improve the process. They agreed to meet for an hour every day for a week and then once a week for an hour to structure a corrective action approach. They each delegated responsibility to a key staff member and invited an operator and an inspector to rotate as members of the committee. Thus began an investigation that revealed a number of things needing correction. The conveyor belt moved too slowly, allowing plates to pile up. This resulted in some plates crushing others. The method of aligning the plaque in the shear was improved through a different guidance system. Different inspectors used different criteria. They were provided with standards to inspect periodically to maintain consistency among themselves and with regard to time elapsed. The tote boxes carrying the plates had radii in the corners that caused damaged plate corners. The fall to the conveyor belt from the shear was too high, so it was reduced. In a matter of 6 weeks the reject rate was down to less than 4%. The five or six rework operators were put onto more productive work. One inspector was removed and placed elsewhere and the second inspector was able to divert some attention to inspection of other products.

What was done here focused the attention of a capable and knowledgeable group of management and operating-level people, encouraging them to analyze a process—an accepted way of doing things—and providing them with a charter to make improvements. They met every day for an hour or less. They used the data collected by the quality control inspectors, analyzed the causes of the problems, and fixed them. They had a higher-level management committee to call upon for the resources necessary to design and have a new cutting alignment tool built and to get replacement tote boxes. Speeding up the conveyor was done simply by turning a dial; raising the conveyor required only simple carpentry work.

What was at work here? A dedicated group of people with the charter and freedom to make positive change, together with a realization by middle management that things could indeed be improved, within an environment that fostered change. The result was not only product improvement, but relocation of at least six people into more productive assignments. It is not enough just to improve. Costs can only be saved by reducing people and material usage or by increasing output with the same personnel and materials.

The events in this example took place in 1957. Today, similar approaches are being taken in many companies. They are often more formalized—perhaps using quality circles, Pareto diagrams, data collection methods, and an environment that attempts to foster change. And they work—sometimes. We cannot accept the status quo as inevitable. We must seek out and improve processes that are operating with defects or whose cycle time can be reduced.

Case History 2: Chemical Concentration of Nitrates

In the processing of nickel–cadmium battery plaques, several processes are critical to the successful performance of the battery. One of these processes is chemical impregnation. This involves immersing batches of plaques into a solution of nickel nitrate or cadmium nitrate. The specifications govern a variety of processing parameters, one of which is the concentration of nickel nitrate. The specification, originally established in the laboratory and confirmed in pilot production, was based on experimental results and the capability to maintain the specified level in the factory.

Over the years as factory production expanded, the process of adjusting the concentration was changed from individual mixture adjustment to a central storage system that dispensed the chemical to the tanks in which the process took place. Due to the large volume of the solution and a centralized system, the capability to control the concen-

Table 1.1 Concentration Level of Nickel Nitrate

	Concentration range				
Exposure Time	*1*	*2*	*3*	*4*	*5*
5 min					
10 min					
15 min					

tration was improved. Several years later when we were trying to raise battery capacity, we decided to evaluate the effect of operating at different levels within the specification. Since there was an opportunity to evaluate the time of impregnation as a part of a statistically designed experiment, an experiment was set up with five ranges of concentration, all within specification limits, and three exposure times. The experimental design looked as shown in Table 1.1. Twenty plaques were processed in each combination, resulting in a total of 300 plaques. There were possible combinations of concentration and time.

The statistical analysis performed on the battery cells that were made from these plaques showed a significant increase in capacity in the fourth concentration level and no difference in capacity for the 5-, 10-, and 15-minute exposure times. Since the standard exposure time was 10 minutes, we were able to introduce a reduced exposure time. This enabled a doubling of production. Since this was not a limitation of production, this production improvement was not really of major economic consequence. On the other hand, the increased battery capacity resulted in a yield improvement from 92% to 96%. Rejects dropped from 8% to 4%, a dramatic change.

These results helped reduce the rework somewhat, but even more important was the step toward achieving 100% yield. This was a major goal because there were 10 operators engaged in testing cells, others used to bolt the cells into series strings for charging and discharging, and operators who then disassembled the cells and reassembled them into battery containers for battery-level testing. For the battery (usually consisting of 24 cells) to pass its specification, all cells had to meet their specification. Eventually, 100% yield was achieved, but it took additional tests using different processing parameters.

The message here is that the mere existence of a process specification should not be regarded as inviolate when seeking improvements on a continuing basis. Thought processes leading to breakthroughs in improved performance demand that specifications be tested for validity on a more-or-less ongoing basis. In this situation the specification level for nickel nitrate concentration was too broad. Narrowing the specification limits was achievable, as the process had been improved, but it took years to recognize that a narrower limit would produce a better product. If the results had shown the concentration increasing from level 1 to level 5 rather than peaking at the fourth level, with the battery capacity continuing to increase, it might have been wise to raise the upper specification limit for concentration. If that had been done, it might have been necessary to evaluate the batteries under other operating modes, such as high-temperature performance, low-temperature performance, shelf life, cycle life, and other performance characteristics. Since the original specification limits were not exceeded, adequate data existed for these other performance modes and additional testing was not needed.

Case History 3: Reuse of Sodium Hydroxide

Other experiments were conducted in an attempt to achieve better yield from sodium hydroxide. This also involved processing of battery plaques through a step called formation. In the formation process, sodium hydroxide was used together with a charging current to change the chemical composition of the nitrates. This process was known to result in degradation of the sodium hydroxide, so the specification limited its use to two cycles, after which the chemical had to be discarded. This process continued for years, and production reached the point where a tanker truck delivered sodium hydroxide three to six times a week.

Questions arose about the effect of using the sodium hydroxide three or more times. There were also questions about the charging current. So it was decided to run an experiment varying both the number of uses of the sodium hydroxide and the charging current. An experiment designed similar to that in Case History 2 was set up. This time, three charging currents were used with four usage cycles. The experimental design is shown in Table 1.2. The same sodium hydroxide was used to process plaques until 16 uses were achieved. The three levels of charging current were used with each combination of sodium hydroxide uses. This resulted in 12 groups of cells. Each group contained 20 cells, which when tested showed slight but statistically significant degradation at 16 uses but no difference at 2, 6, or 10 uses. There were no differences among the charging currents selected, so the standard charging current was retained.

Table 1.2 Reuse of Sodium Hydroxide

	Sodium Hydroxide			
Charging current	*2 Uses*	*6 Uses*	*10 Uses*	*16 Uses*
1				
2				
3				

Since we had no experience wtih cell performance under these processing conditions, cells with 10 uses of sodium hydroxide were submitted to an array of performance tests for assurance that some other cell characteristic did not degrade. Cell performance was as good as the control (2 uses and charging current 2) under all conditions of use. The selection of 10 uses became the new standard and resulted in a direct saving of over $100,000 annually in material procurement costs with an attendant reduction of waste disposal problems.

This is another example of intelligent questioning of established parameters. Very often process specifications are developed under laboratory conditions. They may or may not be checked out in a pilot plant and then are used in production. While in use these specifications should be considered inviolate. But that does not preclude raising questions and using statistical methods for efficient experimentation on process parameters. Factorial designed experiments, Latin squares, or evolutionary operation are powerful techniques that can be used for process optimization. These are discussed further in Chapter 10.

I hope these examples have whetted your appetite to learn more about what can be done to improve quality and productivity in your own operations. The remainder of the book addresses management and technical issues necessary to achieve world-class products and services.

2 · Managing for World-Class Quality

This chapter is essentially a synopsis of the book. It is intended to provide the reader with an outline of what is necessary to manage for world-class quality at a competitive cost. It puts a package of concepts and tools into perspective so that an overview can be developed. Specific subjects are expanded in later chapters. The reader can select subjects where more information is desired and thus pick and choose specific areas of interest. The material in this chapter is a condensation of the lessons I have learned in applying quality methodology to enhance productivity and competitiveness. The specific quality methods are augmented with other material that has been learned or observed in many plants. This material has been further amplified in discussions held over the years with colleagues who were generous with their ideas and recommendations.

World-class quality is essential for those who wish to compete in the domestic and international marketplace. Companies are recognizing that product quality and lengthened warranties are characteristics that sell products. Consumers are flocking to companies whose products excel. Belatedly, American companies have recognized this fact but, unfortunately in many cases, not until large segments of their business have been lost to foreign competition. Winning back some segment of that market and retaining and expanding other product areas demands good product quality. Companies striving to keep costs down are demanding defect-free materials so that inventory costs can be minimized by using just-in-time (JIT) delivery methods. This type of production control requires fully conforming products since products are delivered directly to assembly operations. A product containing nonconforming items results in nonconforming assemblies, which drive up costs and reduce or eliminate incoming validation. The net result provides better products at lower costs, and better product quality sells products, creates more market share, results in higher productivity, and ultimately, provides higher profitability. Managers are now cognizant of the fact that they must improve their product quality if their companies are to compete and survive. Better product quality demands design and manufacturing systems that are compatible.

Quality is no longer simply an inspection function, or even just a corrective action program. Present belief holds that quality must emphasize problem prevention. But I believe that quality must move beyond prevention into creation. It can become an integrated process creating profit, improving productivity, and providing greater market share for the company smart enough to use a quality management system in the proper way. It can also create pride among the work force and a better quality of working life. The ideas in this book are aimed squarely at the changes needed to achieve these objectives.

Twenty-five key points are made below, some of which are expanded in later chapters to provide the depth necessary for understanding and implementation. The object is to provide enough information to enable readers to start the process.

Point 1: Manage your business to generate information as well as products.

Think about how information might change your operations. If information is generated, what is its source? What type of information is needed, and how can it best be put to use? Traditionally, businesses are run by knowing market demands as to the design and manufacture of a product, and by providing controls over costs and schedules. The information discussed here is information about quality. Quality in its many interpretations is discussed in Chapter 3, but for the purposes of this section, quality means *yield*—the percentage of good products of the total being manufactured. Although this may be considered to be conformance to requirements, the aspect of "fitness for use" is not ignored.

The suitability of products to satisfy customers' needs is essential and the specifications must reflect these requirements. Quality also includes the *reliability* of the product—the probability that a product will function properly in its intended environment for the specified (or expected) period of time.

How does the demand for information change the way we operate? Let's examine two scenarios. Under scenario 1, a business is going to be run simply to turn out products. In scenario 2, the business will be run to generate information about yield while turning out products. What is different in the behavior pattern in these two scenarios?

Under scenario 1, as the products are manufactured (and this applies equally to office operations and service business), good items are shipped and bad ones are set aside for repair. This process continues with nonconforming products awaiting rework and repair and then being retested and shipped when needed to meet schedule demands or when it is convenient to do the rework.

In scenario 2, the emphasis is on generating information. Therefore, what must be done is to analyze nonconforming products promptly since they contain information. Two pieces of information are needed. The first is the percentage conforming; the second is the cause of items rejected. If the focus is on correcting problems that are causing the nonconforming products, the causative agents will be removed and product yield will increase. Together with the higher yields goes continuing emphasis on improvement. Thus yields continue to improve. Rework is reduced and the effort spent in analysis, repair, reinspection, and retest is reduced and eventually eliminated. The product stream performs better because it includes fewer repaired items. To realize cost reduction, action must be taken to reassign the operators doing the troubleshooting, repair, and rework activities. Furthermore, material procurement can be reduced because waste is reduced. Meanwhile, the test and inspection department is still removing nonconforming products so that only good items are shipped. As product quality continues to improve, there may be opportunities to reduce the number of inspectors or testers (see point 24). Therefore, scenario 2 will invariably result in better products at lower costs. The upsetting aspect is that most businesses operate under scenario 1.

Another example that illustrate this point relates to field or customer support. Most companies regard the customer service operation as focusing primarily on the correction of problems occurring in the field. Valuable information can be obtained from an analysis of the causes for failure so that designs, materials, parts, or manufacturing processes can be improved to correct a problem. There are many opportunities to develop and use the information available from the conduct of business operations to create knowledge that can be used for improvement. This must be an ongoing process with adjustments being made as necessary. Strategically, to realize benefits, the data must be easy to collect, collate, display, and act on.

During the manufacture of hearing aids, an analysis of 100 field repairs showed that 28 were due to a defective volume control. Engineering analysis showed that the wiper blade was inadequately plated. When a change was made in the wiper blade, these field-related problems disappeared. But special analysis and action were required to achieve this; normally, the action stops with repair of the defect. The extra corrective action taken in this instance saved a considerable amount of money, improved customer reaction to the product, and cost very little to implement. The tasks of information gathering, proper presentation (the right format to provide insight into what is happening), and action are mandatory.

Point 2: Use test and inspection data as a prime source of information.

A normal part of any quality control or assurance function is test or inspection. When a test or inspection operation is performed, its purpose is not only to make sure that good products get through but to generate information for use in improving the process that created the product prior to the point of inspection and test. This sounds simple and rather straightforward and perhaps obvious, but to a great extent, industry has been ignoring the information and using the test and inspection function simply as a filter to make sure that only good items get through. By stressing this approach, there will be continuous pressure for improvement. Do you know what your yields are? What are your top three or four causes for rejection? In many companies inspection and testing are not part of the quality function. There is nothing wrong with that as long as information is collected and used to improve the situation.

Point 3: Establish a system that enables information to communicate intelligence.

If you are going to use inspection and test data as a source of valuable information, the information will have to be conveyed to the user community in a readily understandable form. The recipient of the information will be absorbing the data in many other reports, not just the quality report. What must be kept in mind is the information on whether the product or process is operating well or poorly, and whether it is getting better or worse. The second element requires clear identification of the causes of problems.

Point 4: Make sure that the information is timely, relevant, accurate, and scannable.

The information not only has to be timely and accurate but must be scannable. To satisfy the timeliness constraint, inspecting and testing of products or processes must be done promptly. Weeks-old data are often useless and might just as well remain relegated to oblivion. Such data may even mislead about current problems. If the inspection or testing function is burdened with too much work, the operation must be performed on a last-in, first-out basis to enable data to reflect current operations.

There are many opportunities for the collection of data. Make sure that only the important data are reported. Too many data result in ignoring all the data. So it is important that only relevant data be collected. The need for accuracy is self-evident. Inaccurate data contaminate the entire data package, and even the remaining accurate data then become questionable.

Scannability is a key ingredient. The message that the data are trying to convey must be evident at a glance. Highlighting key points may be necessary. The use of computer-generated reports that provide comprehensive multiple pages of data either will not be read, understood, or will not receive the necessary action. If reports do not generate action, they are not worthwhile generating. You should save the time and effort and do something more useful.

Point 5: Organize for corrective action (quality improvement).

The next issue involves organizing for corrective action. There are many ways of doing this and each organization must mobilize for corrective action in a way that suits the character of the organization and its management structure. Often, the most effective organizational structure for corrective action involves a multifunctional approach. In a factory environment these functions should include quality control, manufacturing, manufacturing engineering, and the design function. Sometimes, the latter can attend on a periodic basis, depending on the corrective measures being discussed. From time to time, people from purchasing, marketing, or human resources may have to participate. This is up to the chairman of the corrective action team to determine, along with any help the team members can offer.

In other types of environment, the corrective action team (or quality improvement team) should include, as a minimum, someone involved in creating the product, someone who uses the product, the system definer, and an evaluator. Others may participate on an ad hoc basis. Although other approaches can work, this has been the most forgiving of the variations in individual competence, organizational change, and interference with other actions taking place in the company.

Another aspect of the corrective action system is that it should meet regularly at a prescribed period of time so that the attendees or their representatives can plan their week or their day around the meeting. The best time to hold a meeting is early in the week, as soon as data are available from the inspection and testing functions. Tuesday morning works well because data are generally available from the prior week's performance. The meeting should be limited to 1 hour, and brief minutes should be issued (see Chapter 8 for more details). There should be more than one corrective action committee. The important thing is to address the problems on a timely basis. It may even be sensible to hold a daily 15-minute standup meeting to discuss corrective action, especially if you are operating in a high-volume environment where processes or products can change rapidly.

Point 6: Use the intelligence you gather to take action.

The next issue is to make sure that corrective action happens. One of the most difficult things to do is to implement the ideas that emerge from corrective action meetings. This is why regular meetings must be held and action items assigned, and why actionees must be present so that the corrective action process belongs to everyone. Chapter 8 covers methods for instituting corrective actions.

Point 7: Make sure that the problem stays fixed.

Now you must ensure that the problem stays fixed. It is very common in industry to find that there are cycles of problem recurrence. The cause may be slightly different, but the result is the same. The only sure way to achieve permanence is to provide periodic checking to make sure that the system that has been installed to correct the process is still functioning properly and that process changes, or some other factor, have not been introduced to create more problems. The cycle of checking should be adjusted so that you continually reduce the amount of labor required for the verification process.

Reasons for problem recurrence include changes in personnel, thus losing an important and perhaps unspecified technique; a change in equipment or in wear and tear on equipment; changes in suppliers; changes in tooling; changes in materials; changes in environment; changes in computers or computer program; and changes in management.

Point 8: Use the knowledge to prevent future problems.

Use the data gained to provide a corporate memory and make the information readily available to new, as well as the more senior design and manufacturing personnel. This type of corporate memory can be very valuable in preventing problems as new products are developed. There are times when such corrections can be introduced into computer-aided engineering, computer-aided design, or computer-aided manufacturing processes so that the computer will provide information to prevent undesirable changes and will not permit marginal operations to emerge. Thus using prior knowledge to prevent future problems can be a major benefit.

Point 9: Never add permanent inspection or testing points to fix a problem.

Corrective action should never involve the addition of permanent inspection or testing points. There are times when an extra inspector or test operation must be added to a process to keep operations moving, but this must never be allowed to become permanent since all it achieves is the addition of cost merely to compensate for process inadequacies. If not carefully watched, these operations become permanent, built-in contributors to cost. Similar operations, such as touch-up after wave soldering, are also to be avoided since that type of activity never seems to disappear. As the processes improve, operators or inspectors will simply become more critical, tending to maintain the defect rate. Imagine an operator reporting that there is no work to do because no touch-up is required. The quality system must be dynamic and requires constant review based on the results of inspection, testing, and customer reaction.

Point 10: Use statistical methods to control processes or achieve permanent improvement.

Many of these statistical methods are available, including frequency distribution analysis, variables and attributes control charts, correlation and regression, Pareto analysis, and statistical design and analysis of experiments. These statistical methods are the most sensitive and cost-effective ways of controlling processes and gathering knowledge in a format enabling problem prevention and rapid problem identification, and helping corrective action. These techniques provide much of the benefit of a statistical quality control program and are discussed in Chapters 7, 9, and 10.

Point 11: Use statistical methods to aid in the continual search for process or product improvements.

More advanced statistical methods can be used to search for process or product improvements. These involve evolutionary operation, response surface methods, experimental design, multiple correlation, analysis of variance, or analysis of means to help improve the product or process in a cost-effective manner. These techniques, and others, enable processing parameters to be optimized so that process capabilities improve. The net result must be a process capable of generating products that are 100% compliant with the product specifications. This procedure may also involve analysis of the specification requirements to determine whether they have been established at the correct levels. It may be desirable, for example, to tighten specifications or shift specifications in order to gain a competitive edge. On the other hand, the specifications may be relaxed in some situations because there are no advantages to be gained by having a specification tighter than the process capability because the user is satisfied, and it is impractical to improve the process.

Point 12: Establish a yield (percent conforming) specification.

Corporate operations must focus on improving the product or process yields because better yields result in improved productivity, overall better product conformance, better field operation, and the attendant growth in market share. One of the problems plaguing American industry today is that management emphasizes schedule and cost. Not only are these items critical, but both schedule and cost can be measured against an established standard, and it is easy to do. It is not uncommon to measure schedule compliance on a daily basis. Costs are measured less frequently—perhaps every week or month. But what about quality reviews? How often can they be held? What are the criteria for yield?

In the absence of a standard, quality has become the residual factor and progress is normally measured against past performance. What must be done is to establish a yield specification—the percent conforming for each inspection and test point. Then measurements can be made against the established yield specification. At this time, few companies have established such a procedure. Experience shows that when the engineering department is asked to establish a value for percent conforming, the normal reaction is that 100% conformance should be the criterion since the design has been made to work on several prototypes. Another ingredient in the design equation is how well the design is centered between the specification limits and how well the process capability compares to the specification limits. The engineers must also learn that while the design affects yield, there are factors working to decrease the yield. They should be responsible for establishing the yield specification. Support should be provided by manufacturing, quality control, and manufacturing engineering personnel.

By establishing such a specification and measuring performance against it, proper attention is devoted by lower-level management to satisfy upper-level management demands. Lower-level supervision responds to upper-level management pressure. When management pressure is on schedule, lower-level supervision provides compliance to schedule; when pressure is on cost, lower-level management provides compliance to cost; and when pressure is on yield, lower-level supervision provides compliance to yield. This process must be ongoing: once a yield specification is reached, the specification should be reestablished.

Point 13: Establish a reliability specification.

The better the yield, the better will be product conformance in the field. This addresses initial performance only; to have a world-class product line, the product must also perform satisfactorily over extended periods of time with little or no maintenance. Therefore, a reliability specification must be established and a product life cycle must be determined. The reliability specification emphasizes long-duration trouble-free performance. There are many techniques for achieving high-reliability products. These are discussed to some extent in Chapter 15. Some of these will result in increased costs and may not be practical from a product point of view. But methods can also be found to lengthen product life and increase the mean time between failures (MTBF) with little or no cost increase. The need for failure-free performance in the customer's hands must be identified and then must become a part of the design and manufacturing focus because this improves customer satisfaction and provides an opportunity to lengthen warranty periods—another opportunity for a competitive edge.

Point 14: Management must ask the right questions.

If quality is truly the leading goal, the company president must ask questions about quality. This means in-house yield and field performance, customer returns and warranty, and other factors related to product performance. If the company president pontificates about the need for high-quality products but fails to follow through in meetings, managers will perceive the president's real objectives and will try to satisfy those objectives rather than the exhortations for quality. Everyone says that top management must be behind any thrust for quality improvement. The role of top management in this arena is to state the objectives and ask the right questions continually and at all levels. Yields and other quality issues must become a regular part of management meetings.

Point 15: Know the cost of quality.

The cost of achieving conforming products that satisfy customer demands must be known. The traditional accounting methods of labor, material, and overhead do not disclose sufficient information to make visible excess costs due to scrap, rework, troubleshooting nonconforming items, touch-up, and other costs incurred because a process is not capable or because inspection or testing discloses deficiencies. Although most accounting systems do have separate accounts for warranty and product recalls, they often do not keep separate records for scrap, rework, reinspection, troubleshooting, customer returns, field costs for servicing nonconforming items, and the income differential if a product is graded as a second. All these costs are considered nonconformance costs. Even rarer are accounting records showing the cost associated with all incoming, in-process, and final inspection and test, regardless of whether the cost is incurred by the quality, manufacturing, or engineering department. Yet if known, these costs may be avoidable. Cost avoidance has many benefits. If good products are produced instead of nonconforming products, productivity is increased, outgoing product quality is improved, and the savings drop directly to the bottom line. If inspection and testing can be reduced, costs not only reflect this reduction, but inspectors and testers can be made into producers, further increasing productivity.

Point 16: Establish a strategic plan for quality.

Quality costs can be significant. It is not uncommon for such costs to range between 15 and 25% of sales—more than most profit margins. If these costs can be reduced and the savings dropped to the bottom line, profit margins can be increased dramatically. With these costs so large, the role that quality occupies in the long-range plans of a company must be considered to be significant. It is therefore appropriate to have a strategic plan for quality to go along with other business strategic plans. The strategic plan for quality can focus on many issues, and measurable goals can be established for 1-, 3-, or 5-year periods. Such issues as reducing warranty costs by 50%, reducing quality costs for nonconformance and product cost by some percentage, improving productivity, increasing yields at any number of key points, and training management and operators in statistical procedures or quality costs are typical established measurable goals. This can be a profitable and worthwhile exercise. If there are strategic plans for marketing, engineering, manufacturing, personnel, and capital investment, a strategic plan for quality—long and short term—is warranted.

Some companies now have established such goals as "a tenfold improvement in 5 years." More aggressive companies have stated that a "100-fold improvement in 4 years" is essential—and achievable. Selecting the proper yardstick is vital in this respect. Motorola advocates the simple concept of "defects per unit" (where a unit is an opportunity for a defect to occur), thus providing a universal measurement standard.

Point 17: Use high technology to improve product and systems.

Although this is a book about quality systems and methodology, it is recognized that managing for world-class quality involves other disciplines. The use of high technology to achieve better products at lower costs is vital to operations. Computer-integrated operations where common data bases are used for the engineering, drafting, manufacture, test, and inspection of products will improve productivity and schedules, and are likely to improve product quality and reduce costs. Furthermore, the product cycle between market research, product design, and release can be substantially reduced, enabling good products to be brought to market more quickly. The use of expert systems can further speed the introduction and simultaneously reduce the likelihood of introducing design or manufacturing weaknesses into new products. Artificial intelligence that utilizes computers to impute logical analysis is just starting to be used.

High technology can also be used for the collection and analysis of data. Automatic test equipment has data logging capabilities. This enables results of tests to be displayed promptly in formats conveying information that can be used by manufacturing, quality, and engineering personnel to correct problems related to particular product part numbers, location, problems, operator deficiencies, other common causes—and this information is free!

Similarly, inspection and test equipment is available that generates data in frequency distribution plots, control chart formats, three-dimensional layouts (to represent the defect location), and other displays. Computer programs are available to optimize processes without the need for a statistician. These are becoming more user friendly with time. The expert systems can create more error-free processes, such as determining machining parameters, speeds, feeds, sequences, and so on.

The availability of custom large-scale integrated circuits (CLSIs) or application-specific integrated circuits (ASICs) using controlled designs and standardized processes can result in less costly products with greater reliability. These technologies continue to expand at a rapid rate. They must be assessed on an ongoing basis to assure that advantage is being taken of current technology. The timing must consider development status of the new technologies to avoid premature application of the product or process before its reliability can be assured.

The entire office automation process must be considered for all businesses, service and product alike. Obviously, the service industry, where the product is often paperwork, can benefit substantially by these advances in word and data processing. But even in manufacturing industries, production planning, purchasing, personnel, sales, and other functions can be greatly accelerated with greater accuracy. However, one must be on guard for defects of a different type. For example, an entire misplaced paragraph may occur with word processing but is unlikely with typewritten copy.

Point 18: Improve your products beyond customer expectations but keep economics in mind.

This point is best illustrated by the Japanese semiconductor companies, which produced the first semiconductors where nonconformance were measured in parts per million rather than percent nonconforming. For years, the electronics industry using semiconductors tried to encourage semiconductor manufacturers to reduce nonconformance levels. Over a period of years, the nonconformance rate as measured by the acceptable quality levels (AQLs) specified were reduced from 1.0 to 0.65% to 0.4 to 0.25% with much antipathy and debate. Then along came the Japanese, who provided defect rates on the order of 100 to 150 defects per million. When you realize that 0.25% defective is equivalent to 2500 defects per million, the magnitude of the problem becomes apparent. Now U.S. companies are competing at the same defect level but have lost much of their market share. This improvement has been achieved through technology improvements and careful attention to process improvements and control and also by making sure that the processes are capable of meeting the specifications. When this condition exists, virtually all products produced will conform to requirements. This eliminates or reduces the need for screening inspection and reduces the need for testing. Getting there is another story—one of the goals of this book is to point you in the right direction.

Point 19: Use your suppliers to support your objectives.

The supplier base available to many companies is not utilized to its fullest extent. There are several ways in which this can be enhanced. First, the supplier should be capable of delivering products sufficiently defect free to enable the receiving company to eliminate incoming quality control. The suppliers should have a sufficiently good quality system to enable them to produce at low reject rates. This means that the supplier is using statistical process controls to control the process. Then as a final operation, the supplier carries out final testing and inspection. Upon receipt, the receiving company then reinspects or retests the product, or both. Why?

If the supplier's job is done properly, the incoming quality control departments that exist in most companies should not be needed. Why should substantial incoming quality control exist? If this could be reduced or eliminated, tremendous savings could result with no decrease in product quality. Companies are now recognizing this fact and are demanding defect-free products in order to use just-in-time inventory and delivery practices. The opportunity exists to do this in many industries.

Another way in which a vendor can help is to identify design parameters that should be used in the product. Vendors have a great deal of experience in designing and manufacturing particular products and that expertise should be put to use. One of the problems in doing this is the reluctance of in-house engineering staffs to use this knowledge.

This attitude must be overcome to take advantage of the knowledge and experience in the vendor's plant.

There are still other opportunities to use your supplier. The advent of CAD/CAM enables data to be electronically downloaded to suppliers at remote locations. In this way, the geometry of the design is provided and suppliers can use their trained personnel to program the tools for their manufacturing operations. General Motors has developed a standard called manufacturing automation protocol (MAP), which a machine shop can use to acquire design data to support GM's needs. This may become an industry standard. Capabilities such as this exist, enabling the smaller manufacturer to do similar things with other CAD/CAM suppliers. Vendors can be a great asset; use them accordingly.

Point 20: Stay close to your customers.

To provide customers with products they can use with satisfaction, you must know what they want. This requires that close contact be maintained with them. It is desirable to have the quality control manager visit customers regularly. It may also be desirable to have the design and manufacturing managers visit customers occasionally to see how products they design or manufacture are used. Some companies even send operators to visit customers to see how their products are used, but this is rarely done on a regular basis. Recent literature discusses at great length the value of keeping in close touch with customers. Other benefits will also accrue. For example, you may obtain information about your competition from your customers. Bear in mind that the customer may be next department, and satisfying the needs of this customer is also mandatory.

Point 21: Know your competition.

Knowing the marketplace and benchmarking the performance of others is a key way to maintain visibility in the marketplace and compete effectively. There are many sources for this information. It is possible to find out a great deal about what the competition is doing from such sources as customers, suppliers, and sales people and by attending conferences and conventions. Published literature is another source of good information about competitive products, as are government agencies.

Point 22: Avoid gimmicks.

The people in the plant know when management is serious about a new system. For example, if there is a decision to introduce quality control circles and management has not taken the time to learn the process or is not openly willing to accept recommendations from factory or office personnel, the process is doomed to failure. QC circles require a participative management style; autocratic managers are simply incompatible with QC circles. The use of posters and slogans trying to get workers to do better without getting management to do so will accomplish little good. The problems start with management; the solutions should also begin there. Do not misunderstand. The presentation to operators of data they can use to determine how they are doing is very valuable and should not be reduced to the category of posters, pins, and T-shirts. Keep in mind the outlook of many workers. Some people will go along with an idea even when they know that management is not serious. They know that with patience, this too will pass and things will return to normal. If you are serious about quality improvement, your actions and words must reflect this fact. This image needs daily reminders. The questions that managers ask at meetings provide excellent and inexpensive reinforcement.

Point 23: Provide an environment in which people can be self-motivating.

Industrial psychologists tell us that motivation must come from within a person—that it is not something that can be induced. The work environment must reflect management's respect for the individual and his or her ideas. To perform up to their capability, people in management as well as operators must feel this respect. There must also be consistency in objectives reflected throughout the organization. When the general manager states that he or she wants good-quality products (conformance) but the foreman tells the operator to hurry up and get a questionable product delivered quickly (schedule) and never mind about quality, the message getting through is that quality does not matter—and this is the message passed along in the cafeteria and in the rest rooms. For this reason, consistency is essential. What gets preached from the pulpit must be reflected in meetings and appear in the performance measurement criteria of supervisors. This is the only way in which people can believe and react to management in the way that management would like. There must also be an environment in which people feel free to express their ideas without retribution. Properly used, this can harness all the resources of a company and create great benefits.

Point 24: Adjust the quality system as improvements are made.

The quality system is dynamic and must be adjusted as product quality improves. When results show yield improvements, inspection and test points should be evaluated to determine whether they can be reduced or eliminated. For every test or inspection person who can be transferred to a production assignment, there is a double benefit. The appraisal cost is reduced and the direct labor base is increased. Productivity is thereby enhanced. Ideally, evaluation functions should be kept to a minimum. If the process is capable of producing defect-free product, inspection and test costs will be minimized. Some process controls are necessary to guard against unforeseen process shifts, and these require adjustment as the situation changes. Inspection and test do not contribute value when the product contains no nonconformities. Thus they can be eliminated or reduced together with other functions that do not contribute value.

By contrast, as quality gets poorer, the quality system should be adjusted upward but this violates point 9. There may be a temporary adjustment upward, but the temporary nature of this must be reviewed to make sure that it does not become a permanent part of the operation.

This discussion presumes that an optimum quality system is in existence at the time improvements are introduced. It is probable that the existing quality system is not optimum and that added process controls, tests, or inspections may be needed. Perhaps more quality engineering is needed. The process of introducing and using the preceding 23 points should result in adjustments to arrive at an optimum level.

Point 25: Adjust personnel, material, or specifications as changes occur.

Adjustments in the quality system are only a part of the changes that must occur as quality levels improve. Along with reductions in defects, the amount of labor used to troubleshoot, repair, and retest rejects must be reduced. Experience shows that companies with rework departments never seem to decrease the personnel in those departments regardless of the amount of yield improvement. As fewer rejects are provided, the time required to repair them simply lengthens. Conscious action must be taken by management to prevent this. If rework is done by the producing operators, the reduced rework rarely results in increased throughput. The next time an operator engaged in touch-up advises management that no touch-up is required will probably be the first time. This is all stated without derogating the work ethic of operating personnel. That is just human nature. The vast majority of the work force wants to do a good job and put in a fair day's work. There is simply a pattern that develops which must be changed to take advantage of improvements.

Examine opportunities to reduce labor in nonproductive areas. An example of this is in the calibration activity. Calibration of measuring tools and equipment is essential. Often, the frequency of calibration is set at a fixed interval. In one factory a computerized system was introduced to determine a *figure of merit*—the percent of time that a particular model of test equipment required no action when calibrated divided by the total times calibration and adjustments were made. Over a period of several years, it was discovered that most equipment was being calibrated too often. Results showed that calibration frequencies could be reduced on one equipment type from every 3 months to every 24 months. Varying levels of improvement were possible on other equipment. The result was that one-third the number of people are now required to handle twice as much equipment. This particular area could also be streamlined by the use of newly available self-calibrating instruments.

The only way that costs can be reduced is through reduced labor and material usage or increasing good product output using the same labor and material. Care must be exercised through labor adjustments that are made by improvements in quality. If people are laid off, there probably will be adverse worker reaction. Other means must be found for labor adjustments until such time as market share increases raise the demand for more labor.

Similarly, with the reduction in defects, the amount of material consumed can be reduced. To support schedules, most companies order excess materials to compensate for existing reject rates and the longer cycle time resulting from in-line deficiencies. As the yield improves, this over-order quantity can and should be reduced. Otherwise, the material costs will remain the same and the excess material will gather dust, take up space, and raise inventory costs. Companies trying to use just-in-time delivery systems recognize this and also realize that products must be essentially defect free for such a system to work effectively.

The final point is that improved processes may enable specifications to be tightened. This should be done only if an economic or marketing advantage is achieved by so doing. Tightened specifications may reduce rejections in future operations. Wouldn't it be nice to tell your customers that you can provide products to tighter specifications than they require at no extra cost?

In summary, if you can improve, make it pay off.

Summary

Using the points in this chapter and other ideas in this book should help make you a more successful competitor in the world marketplace. But situations will continually change, so innovation, dedication, and statistics are essential to leap frogging others and keeping the lead once you have gotten it. John Ruskin, a nineteenth-century English writer, said: "Quality is the result of intelligent effort." That is the message of this book.

3 • What Is Quality?

Introduction

In Chapter 1, world-class quality was defined as a product or service that is the performance and price leader in its field. Therefore, it is capable of capturing and retaining market share in the United States and abroad on the basis of superior performance initially and throughout its useful life. Quality has received much attention recently and will continue to be a major business factor, but even among professionals there is disagreement as to an acceptable definition. Advertisers love to use the word *quality* because it conveys goodness, desirability, and satisfaction. There are people who may be against motherhood, baseball, or apple pie, but nobody is against *quality*.

The dictionary defines *quality* as:

1. A characteristic or attribute of something; property; a feature
2. The natural or essential character of something
3. Excellence; superiority
4. Degree or grade of excellence

Thus quality can have many facets. Some common perceptions of quality are not adequate to enable business to deal with them in a practical sense.

QUALITY IS GOODNESS. This is the general understanding implied when the word *quality* is used, but it is so general that it cannot be used to convey information.

QUALITY IS PERFORMANCE. If a new product performs well, it can be considered a quality product. This performance covers a wide range of parameters and is based, to some extent, on compliance of the product to the advertising, express or implied. This interpretation of quality is more akin to the reliability definition stated later.

QUALITY IS BEAUTY. If a work of art is beautiful, it can be considered to be a "quality" work of art. A piece of lovely crystal has a beauty about it which will cause a perception of quality to accrue the item. This definition is also not useful for a business activity, but it is an important market consideration.

QUALITY IS STYLE. For those products where styling is an important consideration, such as furniture, clothing, and many other products, style is an element to be viewed as a quality characteristic.

QUALITY IS FEATURES. Quality conveys the concept that there are many characteristics of a product or service that exceed those of the competition. In the case of an automobile, power seats, power windows, and power brakes are features not essential to the automobile, but they convey an impression of quality. Features and the service they provide to customers are also important marketing aspects. Features are often a principal selling point.

QUALITY IS CONSISTENCY. One can go to certain fast-food restaurants any place in the country and expect to experience the same type of product regardless of location or time of day, month, or year. The hamburger may not be "gourmet," but you usually know what to expect. This does not mean that the product is inferior—merely that one knows what to expect and that this expectation is fulfilled on a consistent basis.

All of the factors above are related to quality but are difficult to measure and are not normally a part of a professional assessment of quality, although product quality (consistency) is carefully measured in many fast-food chains and in other businesses.

Measurable Quality Definitions

QUALITY IS CONFORMANCE. Many quality practitioners consider conformance to be the essential definition of quality. If a box of cookies is to weigh 12 oz, conformance would be boxes of cookies weighing a minimum of 12 oz. The federal government frowns on underweight packaging and on the companies that violate these standards. On the other hand, boxes weighing more than 12 oz are undesirable from the manufacturing point because more cost is incurred for the higher weights. This is a clear definition and can be used as a criterion against which measurements can be made. This definition overlooks the design suitability of the product since conformance is the sole criterion for acceptance.

QUALITY IS FITNESS FOR USE. This aspect of quality considers not only the performance cited, but the suitability of the product to perform a task as advertised. It therefore takes into consideration the basic design of the product, as well as the conformance of the product to the design. If there is a design inadequacy that causes the product to fail to fulfill its use, the product fails the fitness for use definition. Design, parts and materials, processing, workmanship, and service must all combine into a satisfactory product. A design or process deficiency may be inherent in a large proportion of the product. Product recalls are most often related to such design or process problems rather than workmanship problems. This definition can also be used for measurement purposes providing information in making business decisions. Data on fitness for use must be obtained from the field while data on conformance are available at the place of manufacture.

QUALITY IS RELIABILITY. Reliability is defined as the probability that the product will perform its intended function or mission in the specified environment for a prescribed length of time. In this sense, reliability takes into consideration the domain of time. It must not only satisfy its application when new, but must continue to satisfy its application throughout a vaguely defined period of life. Many products now carry warranties. The length of time specified in the warranty is generally an indicator of the minimum length of time the product can be expected to perform satisfactorily, but customer expectations exceed warranty claims by a considerable amount. For example, television sets frequently carry a 1-year warranty. Expectations and experience show that 5 to 10 years are the norm. This feature can be a major competitive factor, since longer-term warranties are, in fact, beneficial to the consumer. Thus product reliability can be a strong factor in marketing and is, indeed, used by many companies for this purpose.

QUALITY IS YIELD. The percent of products conforming to the specification at each evaluation point is considered the yield. This is one of the more important business definitions because it has a major impact on cost. Product and process yields also affect delivered product performance since there is a correlation between in-plant test yields and field performance. If test yields are poor, field performance will tend to be poor, and if test yields are good, field performance will tend to be good. This definition is one that will play a principal role in further discussions in this book.

QUALITY IS CUSTOMER SATISFACTION. The overriding aspect of quality is the perceived value that the customer receives. The customer is the final determination of quality whether that customer is the one who pays or simply the next person (or group) that uses the product or service to perform a process. Actually, the truly world-class entity provides a product or service beyond customer satisfaction—perhaps to achieve customer delight. This is what differentiates a world—class operation from very good operation.

Summary

The characteristics of quality that we have discussed have a profound effect on the product image and result in a wide-ranging marketing advantage. Products whose quality is perceived as superior attract more customers not only when the initial costs are the same or lower, but even when costs are somewhat higher. Consumers do not want to spend their time and money in maintenance or service. They want and expect the product to continue to function properly for an extended time, and they are often willing to pay for performance. Once a reputation for good performance has been obtained, many benefits accrue. For companies with many products, there is an image carryover. People will buy a variety of products from one company because of the superior performance that they have experienced with other products from that company. Once established, the company can "cash in" on that reputation. However, the reputation must be nourished by continued superior performance of the product or service offered. The marketing strategy can integrate this element with the other less definitive aspects of quality mentioned earlier to capitalize on the total image. Throughout, I attempt to be specific whenever the term *quality* is used so that the reader will understand which of the many definitions is appropriate in the context of the particular passage.

4 · The Quality Function and the Chief Executive Officer

Introduction

There are few, if any, chief executive officers who will not proclaim that their organization's products or services are "the best available" or "satisfy their customer's needs" or "are second to none" in their industry. They establish their support for the quality and durability of the items they supply. Some even go so far as to articulate this policy to subordinates or to customers. There may even be a written corporate quality policy. Unfortunately for most American companies, after stating these grand proclamations, the CEO may not understand what needs to be done to convert these statements to corporate actions enabling their achievement.

This is not the fault of the CEO. He or she knows that quality is an essential ingredient to employees, managers, suppliers, and customers. Chances are that the CEO has had little or no experience in the field of quality, so does not know what to do next. In his book *Theory Z*, William G. Ouichi cites a study conducted by Bouchet of the career paths of the top officers of 50 of the largest American firms over a span of 30 years. On the average, they had worked in less than two different functions, and if they had worked in either finance or personnel, they had typically never worked in any other specialty. The likelihood is therefore very small that the CEO has ever worked in the quality specialty. So why should there by any surprise that the CEO knows little, if anything, about quality or what can be done with the quality function to make substantially greater profits. This chapter provides insight for the CEO and other top corporate officers about how to use the quality function not only to achieve this objective but to increase market share and build customer loyalty to all of the company's products.

Policy Statements

Let's start with the statement of policy establishing a company's position with regard to quality. Usually, the CEO endorses the policy and that is about as far as his or her involvement goes. The CEO almost never reviews quality status in each product line unless major problems arise. In the absence of review, the "body language" the CEO transmits is that he or she is not really interested in the details necessary to establish credibility for his or her pronouncements on quality. The CEO will review marketing plans, advertising plans, expansion programs, capital investment programs, profits—positive or negative—costs and schedules, and major problem areas (e.g., union situations, new competitive products, staffing and similar matters), but almost never reviews quality status. In fact, chances are that the CEO often is not sensitive to the data that provide the useful measures of quality status.

Measures of Performance

It is therefore relevant to consider the nature of these useful measures of quality performance. While some measurements may be unique to certain businesses, others are useful to many enterprises. Consider some of the following as examples:

1. What are the warranty costs as an absolute dollar value, and how do they compare to sales? How do warranty dollars compare with the warranty allowance? What is the warranty allowance with respect to sales? How many items are returned, or how many service calls are required? What is the cost of customer returns or customer complaints? What are the principal causes of customer returns or customer complaints? How large is the service department?

2. Do the financial figures provide information on waste—both scrap and all rework labor? Do scrap costs include only material, or is labor included as well? How much allowance is provided because of in-line rejection? Are purchases increased by 10% (or x%) to allow for line losses? Does the rework labor include engineers' and technicians' time for troubleshooting, analysis, and making changes?
3. If a standard cost system is used, how much allowance is provided for yield (percent rejection)? If the numbers are based on past performance, what yield factors were used? Are these yield factors reasonable? How can they be improved?
4. How do the individual plants or product lines measure their performance of quality? How often are these reviewed, and by whom? Does any action-oriented program take place as a result of these reviews, and who takes them? Under whose direction? Who reviews status and results? Is a yield specification established?
5. How much does inspection and test cost in relation to sales or product cost? Does this include all inspection and testing or only that in the quality department? What about cost for inspection or testing in manufacturing or engineering? Are the costs reasonable? How can reasonableness be determined?
6. How much product is sold as seconds? What is the sales differential between first quality and seconds? What is the annual cost?
7. During design, are requirements specified for yield? How much do these yield numbers add to product cost? What is the impact if defects get into customers' hands? What is the industry defect level? What is the yield performance against specified yields? How are yields established?

There are undoubtedly many more questions to be asked, but these will provide a good start. There are many questions that will be unanswerable at the outset, but this should not deter the CEO from pursuing the answers diligently.

Data on Quality

One good source of information about product quality is provided by marketing personnel. Their contact with customers can provide valuable insights into how customers perceive the product. While their perspective may be biased, it is an excellent information source—because good product quality helps sell products. It makes marketing and sales jobs much easier and contributes to better market penetration, market share, and customer loyalty. Competition is based on product features, style, performance, durability, and freedom from repairs. The latter three items are subsets of quality.

In an accounting system structured for visibility into "quality costs," accounting records provide a second source of information. These can reflect excess costs for material or labor and can serve as an indicator of areas needing improvement. More detail is provided in Chapter 5.

Another good source of information is provided by internal test and inspection data. These data reflect in-house product performance and provide a measure of the percentage of a product that meets the established criteria. The availability of this information is vital to the conduct of the business and is discussed in detail in Chapter 6.

The Organization Structure

One of the principal elements in developing a quality (and therefore, productivity)-oriented operation is the attitude of people, management, and labor. Assuming that the message that product quality is an important corporate strategic consideration does get to the troops, how will they respond? People are very quick to perceive hidden messages. If there is no follow-through, there will be no genuine effort to achieve high product or service quality. There will be the routine performance and attention to quality, but no real achievements or breakthroughs will result.

One way that the hidden message gets telegraphed instantly throughout a company is through the organization chart. At the division or plant level, does the vice president or general manager have one staff member whose primary responsibility is quality? If the answer is yes, is this person's title the same as that of others on the staff who have similar responsibilities? A general manager who has a director of manufacturing, a director of marketing, and directors of everything else, but only a manager of quality or product assurance, is loudly proclaiming a lack of real interest. If the salary grades are not identical, the manager is again portraying a lack of real interest, but not as loudly. If the quality manager does not have an equal chance to accede to the boss's job, there is subtle discrimination, and this is perceived.

This raises an issue as to the type of person selected as the quality manager. If he or she is either a misfit or someone who has not been able to make it in another occupation, he or she is the wrong person for the job. Choosing such a person is yet another signal to the plant population that any quality commitment is only superficial. The selection process is an important one because the person must have a good technical background (in statistical methods and product knowledge) and have demonstrated the ability to manage well. He or she should be promotable beyond the quality director to general manager or to a higher executive position. The person must understand the quality role and the management role. There are few higher institutions of learning that offer a degree in quality control. As of this writing, few of the business schools training future top managers offer courses or case studies on the impact of product quality on productivity, cost, sales, and market share. Thus, while a person must have a good education, it need not be in quality or even in statistics. Rather, he or she must have an awareness of the use of data to manage and to improve operations.

In the *Harvard Business Review* of March–April 1974, Schoffler, Buzzell, and Havey of the Strategic Planning Institute point out the importance of quality on profitability and market share in their article "Impact of Strategic Planning on Profit Performance." These data are based on a study of 57 corporations with 620 diverse business. The data have been reinforced since that time.

Getting the Message Across

Let us assume that the quality management function is staffed by a fully qualified person who has equal access to the boss with the other staff managers. What, then, is the next message to convey, and how should this be done? The message is that top management is seriously interested in product quality. One good technique is through the mechanism of reviews. All sorts of status reviews take place at many levels in the business unit. The CEO should include in reviews the subject of quality, not just in the sense of examining the major problems. Questions should be directed at the cost for lack of quality, yields, and customer assessments.

The cost questions should consider warranty, customer returns, and field costs due to poor product quality on a product-line basis. There should also be reports on in-house failure costs associated with troubleshooting, scrap, and rework. In these areas, the accounting department should have figures and the CEO should be presented with absolute levels as well as trends. For a small business, the personal computer has made gathering data and presenting information relatively easy provided that the system is established to make possible proper cost classification.

Product-line managers should be required to establish goals for the cost of quality, and these should be reviewed and approved. If the costs are too high, questions should be raised and answered. Do not forget: All the costs mentioned above are excess costs that reduce the bottom line; they cut directly into profits.

These kinds of questions raise important economic issues. If allowance are made for lack of quality, it is vital to know why. What is necessary to reduce these costs? It was this type of thinking, not QC circles, that led Japanese manufacturers to surpass U.S. manufacturers in product quality. It began with management involvement and an understanding of where excess costs due to poor quality could be eliminated.

A further area of top management involvement is that of assuring high product quality simultaneously with low cost in product-line yields. How high a percentage of rejected material is being experienced—and if too high, what are several leading reasons for the defects? It is surprising that most top managers are unaware of what their product yields are running. Even more surprising is the fact that some product yields consistently run at a 40 to 60% level. Still more surprising is the attitude of many managers that those levels are in the nature of the process and that nothing can be done. Recall the case study on yields of plates in Chapter 1. Although this situation may be true in some instances, there is usually some action that can be taken to make major yield improvements. For example, a major contribution to product yield or process improvement can be made by organizing to correct problems. This can be accomplished through teams that use statistical process control and statistically designed experiments for problem solving. These subjects are discussed in Chapters 7, 8, and 10.

Once again, using the Japanese as an example, U.S. electronic parts manufacturers had insisted on using statistical sampling as the means of accepting lots of products. Typical values of 0.4, 0.65, or 1% AQL (acceptable quality level), the most common measure applied to these sampling plans, were used. This equates to products with these defect levels having a high probability of acceptance (on the order of 95%). The Japanese decided that these defect levels were intolerable, and they promptly established new levels. It is not uncommon to find Japanese electronic components in the range of 50 to 100 defects per million. A defect rate of 0.4% is equivalent to 4000 defects per million. Compare the results: 100 defects per million versus 4000 defects per million. Even when a U.S. manufacturer was running at one-tenth of the required AQL level, they were still well above the Japanese levels. Now

U.S. manufacturers are beginning to find ways to compete. They were forced into it by the competition, not by their customers! Industries were slow to recognize these new levels of quality. Some have been so severely affected that they will probably never recover. Think how much better off the American economy would be if defect levels were at 50 or 100 parts per million. Market share would almost certainly be higher, and foreign competition might still be struggling to catch up. Management must act now to prevent the competition from picking them off, one by one, like ducks in a shooting gallery!

Where Do We Go From Here?

It is recognized that top management can not be everywhere and do everything—although top management must do something in the quality area. The starting point is for the CEO to start asking some of the right questions, and ask them regularly. This starts to get the message to the right people. Top management should not have to take direct action to effect change. This is middle management's job. Top management must ask the right questions to stimulate action on the part of a busy middle management. A reorientation of priorities is in order. Management and supervisory measurement and rework criteria must include performance against quality goals. A renewed emphasis on quality to balance its importance against schedule and cost must be developed. The person with the vision and the authority to do so, the CEO, must become active in the process.

This management process must be followed through in product or service development. Quality aspects as well as design and operating features must be considered for new products. Later chapters are devoted to how to achieve a cost-effective organization to provide a product or service that will contribute to the development of profitable business growth.

5 • Better Products at Lower Costs

Introduction

A vital aspect of managing for world-class quality lies in the arena of costs. The title of this chapter may seem to be an anomaly, but it is factual—as you will see. Properly handled, simple cost analyses can identify the location of unnecessary losses that drain profits from operations.

There was a time when there was extensive media coverage of consumer advocates who criticized American industry for laxity in controlling product quality. The results were changes in laws, product design, and manufacturing cycles to reduce the number of recalls being experienced. These were costly and damaging to the company involved and in some instances to an entire industry. Indeed, some product recalls in the food industry caused the companies affected to go out of business quickly. In other situations, companies slowly lost business to tougher competitors and merged, failed, or suffered a permanent loss of market share.

Management perspective toward product quality has now been altered by the reality of competition. There are still industry critics demanding better product performance, but the driving force has become worldwide competition. Companies recognize that product quality is a strategic business issue that must be addressed initially and throughout the product planning, production, sales, and service cycle. Service industry companies have begun to recognize that better, error-free performance is a key to improved sales and profitability.

Consumer expectations have risen. The consumer is demanding more and can get it. This consumer may be another company, an individual, or simply the next department. Better products and services can be obtained. It is the total life-cycle cost of the product that is important. In essence, this requires a product to work satisfactorily with a minimum of service throughout its useful life. It is industry's job to produce a product or service that meets this consumer demand at a competitive price. The challenge we face is how to accomplish this task. In this chapter we discuss the cost aspect of performance in the manufacture of the product or the fulfillment of the service.

Identifying the Problem

One problem to be solved is how to measure performance. Measures such as percent defective, warranty costs, percent returns, number of service calls, or scrap are often used as criteria in judging quality. When measuring product cost, the usual financial analyses involve material, labor, and overhead, as well as analysis of fixed and variable cost. These measurement techniques are no longer adequate and are being replaced. Only after competitive pressures forced quality levels to the part-per-million level did U.S. manufacturers find that they, too, could compete at these rigorous quality levels—too late to retain their market share. Now domestic firms are targeting yield levels of 3 to 4 parts-per-million as necessary to compete in the future—and the target is moving. In the first place, the various quality measures do not necessarily reflect the total picture. Second, in the financial analysis, when considering material costs, allowances are included in the estimates for a certain amount of defective product. Third, labor costs include allowances for rework, retest, and reinspection without segregating these from the basic labor costs. Since rework, retest, and reinspection can be eliminated, these costs must be segregated to provide visibility into possible savings.

Finally, it is not uncommon to include added operations to correct inadequate performance. For example, it is common to include touch-up operations after an automatic wave soldering process because the solder joints contain too many nonconformances. This is recognition that the initial operation was done inadequately. The touch-up costs should be considered as over the basic production cost, but this is rarely identified. These practices add to cost without corresponding value.

By combining the quality aspects and cost reports into a single quality cost analysis, these shortcomings can be remedied. A different perspective can be gained by management and alternative courses of action can be identified. It is essential to structure the system to provide visibility that can lead to action.

Quality Cost Elements

Ideally, the costing system should be jointly designed by the finance and quality departments to provide the basic cost elements necessary to structure a quality cost report. Most of the time, the cost structure does not provide sufficient information on these cost elements, and modifying the cost system may be difficult. In the past it has not been easy to obtain finance department support, but competition has assured the cooperation of the financial community. In some situations, some estimates may have to be made by quality control personnel. Although undesirable, it may be necessary to get started under these circumstances. Using a few of the many possible cost elements can provide a basis for insight and may enable better decision making.

The four basic quality cost categories are prevention, appraisal, internal failure, and external failure. *Prevention costs* are those associated with efforts to prevent defects from occurring. *Appraisal costs* are those associated with product or process evaluations to determine conformance to specifications or requirements. *Internal failure costs* are those associated with correcting a product or service found to be defective or substandard before it leaves the facility. *External failure costs* are those associated with remedying unsatisfactory field performance of a product or service.

Accounting systems rarely identify these key cost elements and even less frequently provide the data in a format enabling intelligence to be gathered. Some typical cost elements are listed below, with definitions provided in Appendix A at the end of the chapter.

Prevention
- Quality planning
- Training and motivation

Appraisal
- All inspection: incoming, in-process, and final

Prevention (continued)
- Process planning
- Vendor selection
- Design review
- Parts selection
- Qualification
- Reliability analysis

Internal Failure
- Rework
- Retest
- Scrap
- Troubleshooting
- Failure analysis

Extra Operations
(designed to apply corrections to other operations)
- Corrective action costs
- Excess inventory costs

Appraisal (continued)
- All test: incoming, in-process, and final
- Quality audit
- Calibration

External Failure
- Warranty
- Customer complaints
- Customer returns
- Added service required to correct product or services in the field (including recalls)
- Seconds

Although there are many other quality cost elements, too much detail burdens the collection and analysis process unnecessarily. However, if you feel compelled to add a cost element that is important to your organization or delete one that is not, it is entirely appropriate to do so.

Note that these cost elements do not include intangible items such as goodwill or buyer attitudes toward the product. Although this is a very real cost of quality (and may exceed all other quality costs in magnitude), it is not included because it cannot be determined with any precision. Management must be aware that these costs exist and, as a result, the external failure costs are often underestimated. Nevertheless, the information that can be obtained is useful in determining new courses of action.

When accounting records do not provide the information in this form, an acceptable procedure to start with is to identify the various elements of the quality cost system from several sources. These sources include estimates from departmental records and from existing accounting records. When presented in this fashion, it is easier to grasp the concept. The data must be sufficiently precise to enable the identification of areas needing management attention. If the combined costs are within 10% of their true value, they can still be useful in aiding management to identify areas needing attention. This awareness may help get the accounting system modified to provide costs in this more useful format. The estimating approach is useful only for starting the process.

The Magnitude of Quality Costs

A consideration in determining the extent to which these costs should be employed is the actual magnitude of such costs. A convenient frame of reference is to compare the quality cost elements with the total sales volume of the company. Although actual values are extremely difficult to come by, due to the natural reluctance of companies to disclose confidential information, reports have been presented indicating a range of total quality costs to be up to 30% of sales, with the internal and external failure costs comprising upward of 50% of these. This results in a total possible cost avoidance as high as 15 or 20% of sales. When this is compared to return on investment, it is evident that a significant increase in profitability can result if a portion of the failure costs can be eliminated. As failure costs decline, it may be possible to reduce appraisal costs as well, thus resulting in further cost reduction.

Experience in companies initiating a quality cost analysis shows that about half the failure costs can be eliminated in the first year of operation. This may be equivalent to as much as 7 to 8% of sales! These may not just get added to profit; rather, a significant improvement is distinctly possible. It is important to develop a realistic estimate of the cost required to reduce the failure costs. To improve a process may require additional design, quality, or manufacturing engineers; introduction of a design review program; initiation of a training program; or capital investment. On the other hand, it is often possible to reduce failure costs with a minimum of investment just by changing management priorities. Remember the example of battery plates in Chapter 1. Analyzing the distribution of the quality cost elements is necessary to determine the type of corrective or preventive action.

What should be the magnitude of costs? Many factors affect cost: type of product, stage of product maturity, product complexity, state of the process development, manufacturing facilities, and other aspects. So it is difficult to state what the magnitude of these costs should be. It is safe to say that they should always be lower than they are. In absolute value total quality costs should not exceed 1 to 3% of sales, depending on the circumstances. Higher values than these offer great opportunity for improving the bottom line. With time, these numbers can be reduced further.

Although the sales base provides excellent visibility, there are situations where another base is appropriate. For example, production cost may be used as the base when the cost program is applied to one department in which material is transferred to another operation with no profit or general and administrative costs applied. Still other bases, such as direct labor or number of units produced, might be used. The point is that there must be a basis for comparison.

Experience has shown that the arrangement of the cost elements in the composite layout is often surprising to management. It is not uncommon to find penetrating questions raised regarding the validity and accuracy of the estimates made to supplement accounting records. In beginning a program, it is essential to make a realistic appraisal of the costs and the investment necessary to reduce them. That is why it is best to enlist the aid of the financial group. A representative display of quality cost data is shown in Table 5.1.

Case History 1: Manufacturing Printed Circuit Boards

In a particular facility manufacturing printed circuit boards, an analysis showed the cost distribution in Table 5.2 and Fig. 5.2. After the first month's data were gathered, it was evident from the magnitude of the external failure costs that there were field problems and that customers were not satisfied with the product. Although these complaints had been registered earlier, it was not until the magnitude of the returns were presented that a course of action was determined. What was the nature of the problem? Why weren't the problems found in-house rather than in the field? To make progress it was essential to resolve these questions.

When the problems were analyzed it was evident that in-house blemishes rejected by customers were not considered defects. Discussions were held with several customers and it was agreed that in-house quality control personnel would add the offending characteristics to the in-house inspection criteria. But even more important, some items that customers considered unsatisfactory were found to be usable without special treatment at the customers' facilities, so some customer rejection criteria were modified. As a result, internal failure costs rose but external failure costs were reduced even more. These results showed quickly in succeeding months.

Although profits were improved, there are still opportunities for major improvements, but these will take longer to develop. Further reductions in internal and external failure costs will require the elimination of defects. After this is achieved, some inspection and test costs may also be eliminated.

Case History 2: Flexible Film Manufacture

In the manufacture of flexible packaging film a special quality cost analysis was performed. Results were surprising to management in that

Table 5.1 Quality Cost Sample Data (Thousands of Dollars)

		Prevention		*Appraisal*		*Internal failure*		*External failure*		*Total*	
Month	*Sales*	*Cost*	*Percent of sales*	*Cost*	*Percent of sales*	*Cost*	*Percent of sales*	*Cost*	*Percent of sales*	*Cost*	*Percent of sales*
March	$5134	$20.5	0.4	$364.6	7.1	$431.2	8.4	$190.0	3.7	$1006.2	19.6
April	6217	31.1	0.5	453.8	7.3	534.7	8.6	192.7	3.1	1212.3	19.5
May	6110	24.4	0.4	409.4	6.7	488.8	8.0	201.6	3.3	1124.2	18.4
June	8229	41.1	0.5	436.1	5.3	510.2	6.2	246.9	3.0	1234.3	14.9
July	7897	47.9	0.6	407.3	5.1	463.2	5.8	231.6	2.9	115.0	14.4

Table 5.2 Printed Circuit Board Quality Cost Data

	July 1984		*Aug. 1984*		*Sept. 1984*	
	Cost	*Percent of sales*	*Cost*	*Percent of sales*	*Cost*	*Percent of sales*
Prevention	$ 780	0.5	$ 848	0.4	$ 2,366	1.3
Appraisal	10,140	6.5	11,236	5.3	9,464	5.2
Internal failure	7,800	5.0	13,144	6.2	11,466	6.3
External failure	15,600	10.0	12,932	6.1	5,824	3.2
Total	$ 34,320	22%	$ 38,160	18%	$ 29,120	16%
Monthly sales	$156,000		$212,000		$182,000	
Cost of sales	134,160		178,080		149,240	
Net profit	21,840		33,920		32,760	

total quality costs were found to be slightly in excess of 27% of sales. A breakdown of these costs among the four major quality cost elements showed that external failure accounted for 25.4% of sales. Further analysis showed that a major customer was unhappy with the effect known as "orange peel." After threatening to return shipment after shipment, the marketing department provided a 25% rebate. This practice became somewhat routine and had been going on for 19 months. When this information became visible, a study team was assembled and after negotiating an acceptable level of orange peel with the customer, the team experimented with some process variables, achieved an acceptable level of orange peel, and the practice became history. Quality costs plummeted to 3% of sales. Even more important, the customer had been seeking another source, and based on the improved product, coupled with the price reduction of 10 to 12% that became possible because of improved yields, decided to stay with the present supplier. So the product was improved and profits were increased. More important: Retaining this major customer allowed the company to stay in business.

Standard Costs versus Quality Costs

Some companies use a standard cost procedure for estimating product cost. The establishment of standards is usually based on past experience. These experiences include yield factors, so the standards automatically include a certain amount of failure cost. When standard costs are the base, it is vital to know the yield figures used in the calculation. For example, if yield is 80%, then 20% of the product requires rework or scrap. These costs are added into the standards. One of every five pieces made is either scrapped or requires additional analysis, repair, reinspection, and retest. A reasonable goal for annual improvement is often set at 10% cost reduction. But the standard includes the 20% rejection figure! So, in reality, the target is very small compared to the potential savings from reducing the failure costs. What is really bad about this is that the company managers think they are doing great when a 10% improvement in standard costs is achieved. In reality, a 10% cost improvement may mean less than a 2% yield improvement. The 20% failures should be attacked vigorously, not tentatively. What is needed is a shovel, not a pair of tweezers.

Actually, the standards need reexamination if a workable quality cost system is to be employed. Such standard cost systems tend to lull management into a false sense of security as long as standards are being met. As soon as competition achieves a breakthrough and selling prices are reduced, the profit margin disappears and new standards get established, too late.

Quality Cost Reporting and Analysis

Some quality practitioners advocate quarterly or annual reporting of quality costs as adequate for control. That frequency may be acceptable after a quality cost system has been in effect for a number of years, but initially, a monthly analysis is needed to effect satisfactory remedial action. Once the fat has been wrung out, it is time to consider reducing the report to a quarterly basis. The quality cost format is important and should provide summaries of costs in the four major areas: prevention, appraisal, internal failure, and external failure with backup details by cost element.

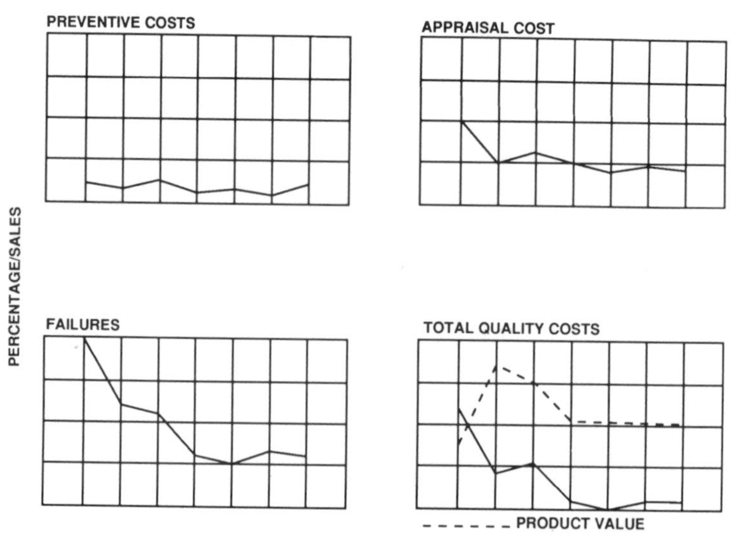

Figure 5.1 Plot of typical quality cost data.

Each major category should be plotted against an appropriate base so that analysis can be performed and action can be determined. See Fig. 5.1 for a representation of these plots. Internal and external failure costs are combined in this case. In giving consideration to the types of action needed, an analysis must be made of the ratio of the major cost elements to the base and their relationship to each other. There is no standard that can be used for comparison. Each case must be decided on its relative merits: type of product, development status of the product, maturity of product, and other factors that affect the quality cost picture. If prevention costs are indeed low, an expenditure in this area will not result in a reduction in other areas unless the expenditure is made for the correct reason.

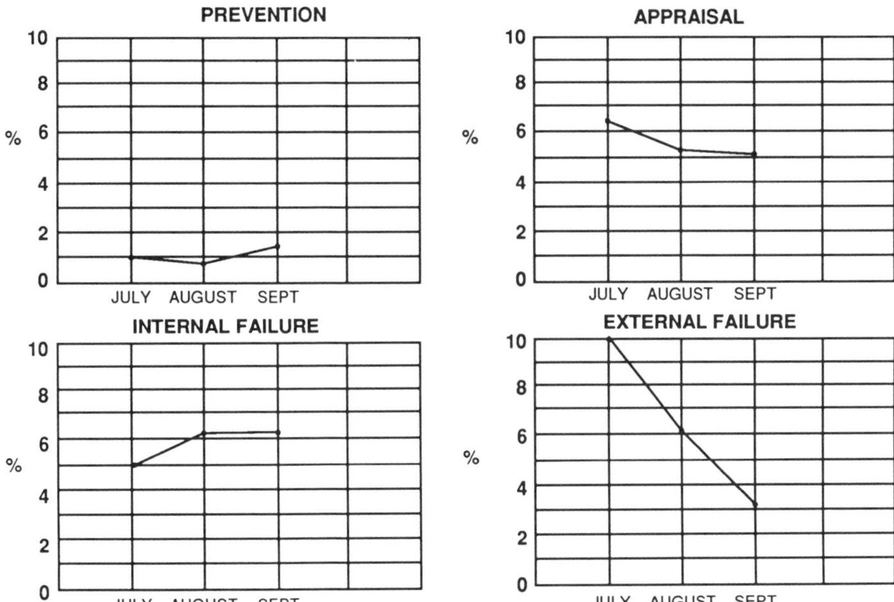

Figure 5.2 Plot of data in Table 5.2 showing dramatic decrease in external failure costs due to effective corrective action.

Plots of quality costs can provide different perspectives as to where problems may exist. Consider the sets of charts shown in Figures 5.3, 5.4, 5.5, and 5.6 together with their respective analyses. In Fig. 5.3 appraisal costs are high, internal failure costs are low, and external failure costs are high. What could cause this? A number of possibilities exist. One such possibility is that inspection and testing are not finding the problems that are causing field rejects. Another is that the repair and retest work is shoddy, allowing defects to get into the field. Still another might be poor design. You can probably develop other scenarios. It is necessary to know how many of what type of nonconformances are found at inspection, test, failure analysis, and field operations. This added insight is needed to help focus on the problem and to speed corrective action.

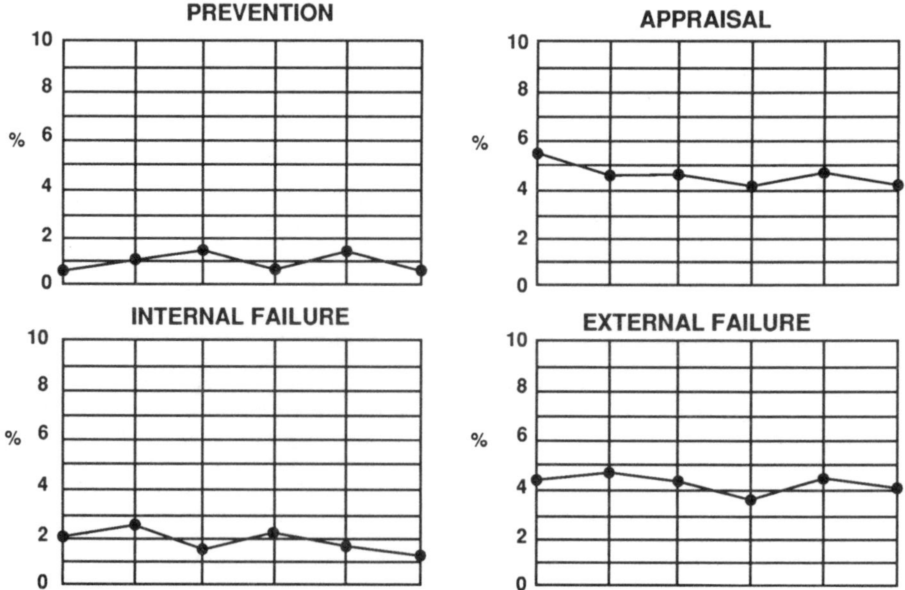

Figure 5.3 Graph of quality costs showing high appraisal costs, low internal failure costs, and high external failure costs.

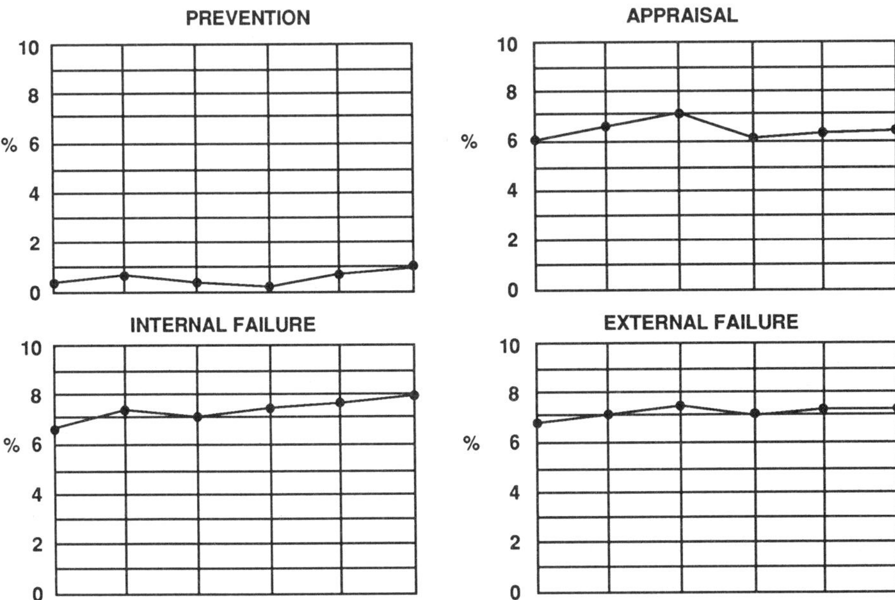

Figure 5.4 Graph of quality costs showing high appraisal costs, high internal failure costs, and high external failure costs.

In Fig. 5.4 we have a true disaster. Appraisal, internal failure, and external failure are all high. It could be bad design, poor workmanship, or a terrible process problem, especially if causes of the nonconformances are the same in each quality cost area. It looks as though this company is trying to inspect quality into its product—and is failing. The nature of the nonconformances will help guide the corrective action process. This is an emergency and demands immediate action. The signal for the problem is the cost, and if a quality cost system does not exist, it is likely that the company will go blithely on its way, ignorant of this drain on profits.

In Fig. 5.5 appraisal costs are high and internal and external failure costs are low. This company appears to be ripe for reducing inspection and testing. Examine the process to see where appraisal costs can be removed without degrading product quality.

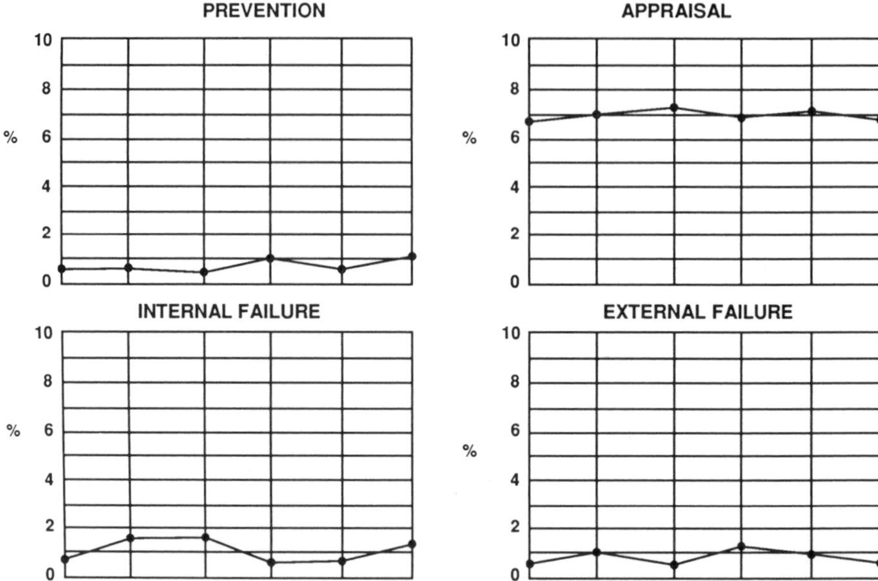

Figure 5.5 Graph of quality costs showing high appraisal costs, low internal failure costs, and low external failure costs.

Figure 5.6 presents us with another situation: high appraisal and high internal failures with low external failures. Are we looking for something the customer does not care about, or are we really finding the defects and removing them? Examination of the defect types and perhaps some market survey is necessary.

In each of the instances there is more than one possible answer. Having knowledge of these costs will help focus attention on the real problem and get it corrected faster. In another instance, an analysis may reveal high internal failure costs and high external failure costs. ("High" costs are relative.) In this situation it is necessary to perform a defect analysis. It is an accepted and realistic conclusion that defect types are not randomly distributed. In every instance there are dominant defect categories and subsidiary categories dubbed the "vital few and the useful [formerly, trivial] many" by J. M. Juran, an acknowledged leader in the field (see Juran, 1988). Determining the nature of the vital few is essential to taking effective corrective action, but lesser problems should not be overlooked.

If the type of deficiency experienced in the field is the same as that experienced in the factory, it could mean that the test or inspection

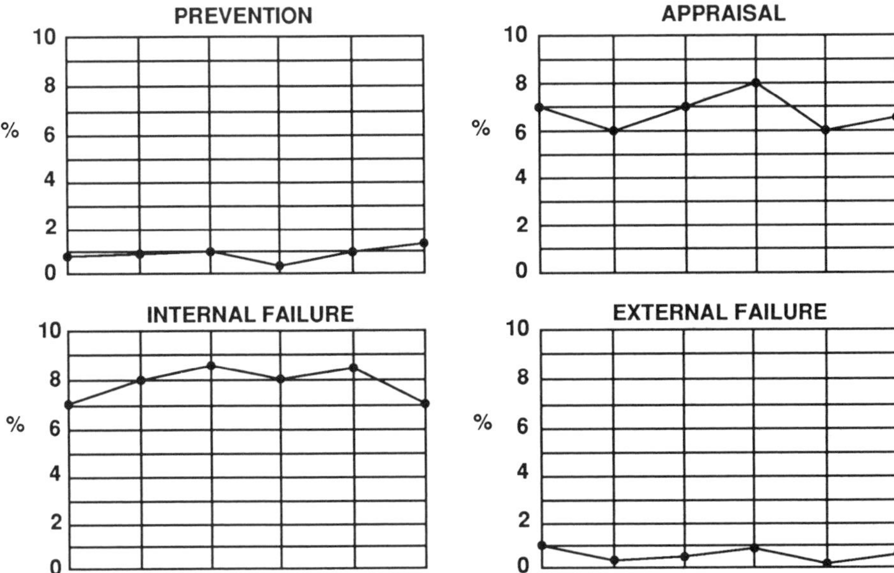

Figure 5.6 Graph of quality costs showing high appraisal costs, high internal failure costs, and low external failure costs.

program is inadequate. If the field defects are different from those experienced in the plant, it might indicate that the proper types of internal evaluation are not being made. In both instances, the causes for the defects must be identified and eliminated.

Depending on the specific nature of the problem, it may be smarter to work on problems that are relatively easy to resolve rather than the most significant problem. This helps to build confidence among problem-solving groups. More difficult problems can be addressed later. This type of action could result in a quick reduction of internal and external failure costs. The strategy that must be used is to identify the type and magnitude of the problems, then set up a program to eliminate them. Properly done, costs will decline and quality will improve.

In all instances where failure costs are reduced, the direct effect is quality improvement. In those instances where appraisal costs decline there is almost always no adverse impact on quality levels. Failure costs are usually far more significant than either prevention or appraisal costs, so that area should receive the focus of attention. If this is done properly, the company will experience a redirection of effort that helps to achieve its cost, schedule and quality goals simultaneously.

Schedules are more readily met when all of the work goes into producing good products.

The Time Effect

The total impact of reducing quality costs over a period of time can be enormous. Consider the hypothetical case of the Ajax Company, which has a fixed sales volume of $10,000,000 for a 5-year period. Profits are estimated at 5% and quality costs at 15% of sales. If an assumption is made that only 10% of the quality costs can be saved annually, the profit figure will double shortly after the third year! This is summarized in Table 5.3 and displayed graphically in Fig. 5.7.

Other Applications

This type of cost analysis has proven useful in other applications. In the justification for new equipment, production improvements or labor savings are usually balanced against investment. The cost of repair, rework, reinspection, and scrap should also be considered. The savings attained through new equipment should be included in the total savings. This will provide more realistic supporting data for replacing equipment. Another use is in the adjustment allowance for procurement quantities of materials. Added quantities are usually procured to allow for in-line rejects and more work-in-process inventory. Declining quality costs are a sure indication that a downward adjustment should be made in material allowances.

Table 5.3 Ajax Company Data

Net sales: $10,000,000
Profits: 500,000
Quality costs at 15%: $1,500,000

	1980	*1981*	*1982*	*1983*	*1984*
Net sales	$10 M	$10 M	$10 M	$10 M	$10 M
Profits	0.5 M	0.5 M	0.5 M	0.5M	0.5 M
Quality costs	1.5 M	1.5 M	1.5 M	1.5M	1.5 M
Savings	0.15M	0.3 M	0.45M	0.6M	0.75M[a]
Cumulative savings	0.15M	0.45M	0.9 M	1.5M	2.25M

[a] If quality costs were reduced 10% each year.

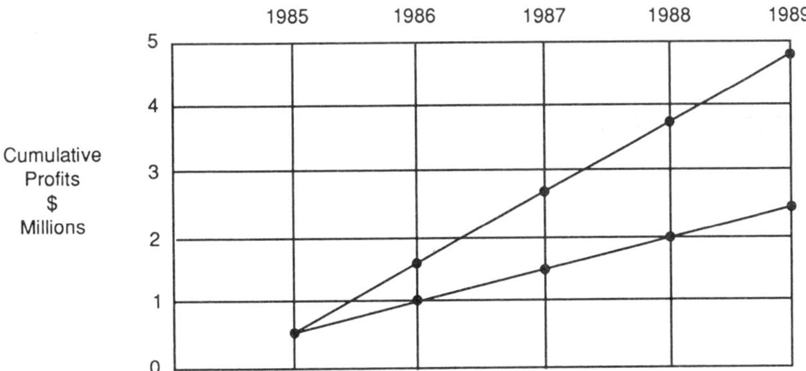

Figure 5.7 Profit growth as the result of reducing cost of quality.

One significant cost not usually included in a quality cost analysis is the cost of extra inventory. Just-in-time (JIT) inventory control does not tolerate defective products. JIT fails when even small percentages of nonconforming products enter the process. Defect levels in the part-per-million category are needed for JIT to work successfully. Rejection rates at various points in the product flow necessitate larger inventories for additional work in process and for added parts.

Figures 5.8 through 5.14 depict how rejections increase the production cycle, inflate inventory, and add cost to product. Figure 5.8 is a typical flowchart used by industrial engineering to illustrate an assembly process. If some rejects are incurred at each inspection or test point, they either get returned to the operator for correction or go to a rework area for analysis and repair. This is shown in Fig. 5.9.

Since there are instances where rework can require products to go back through several earlier manufacturing steps, true product flow may be as illustrated in Figs. 5.10 and 5.11. In Fig. 5.12 three sub-

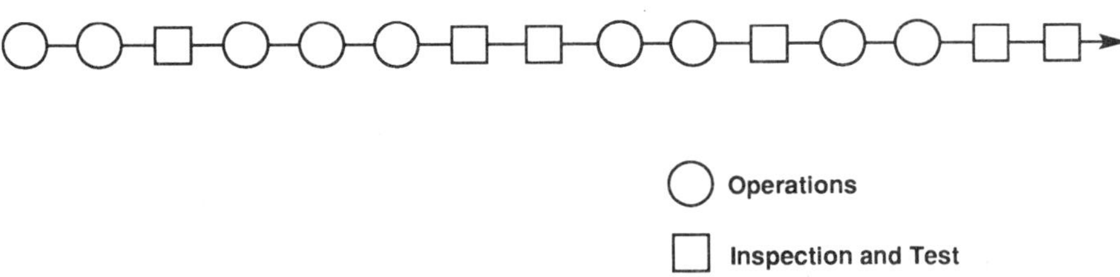

Figure 5.8 Typical flowchart as depicted by production engineering.

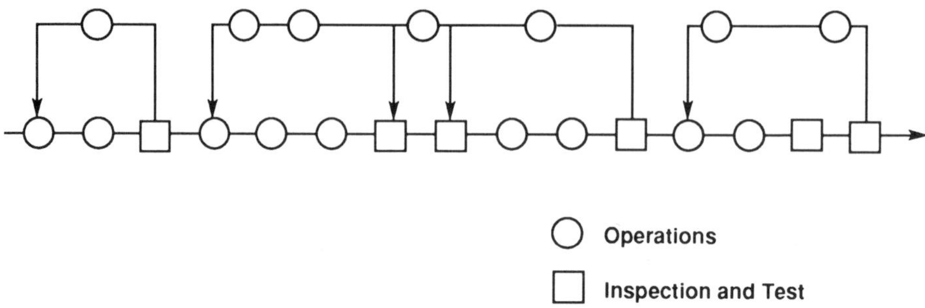

Figure 5.9 Flowchart showing rejection and rework. Each rejection results in a rework batch or product.

assemblies are assembled into a final product. All the attendant rework, reinspection, and retest, and the associated paperwork, add material and labor to the cost of the product and lengthen the production cycle, thereby increasing work-in-process inventory and inventory costs. At the same time, the product quality is degraded through added rework operations and the resultant outgoing product is not as good as it would have been had fewer (or no) nonconformances existed. It should be evident that the schedule time is also increased as a result

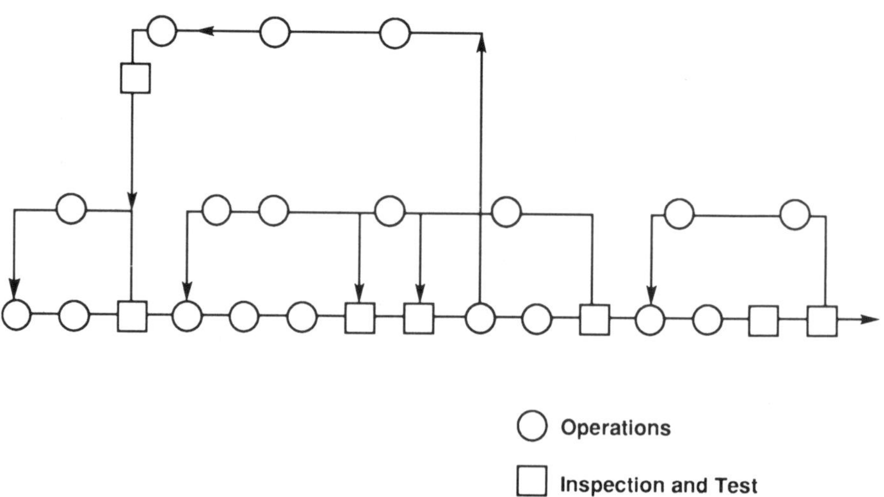

Figure 5.10 Additional rework operations sometimes result in backtracking more than one position.

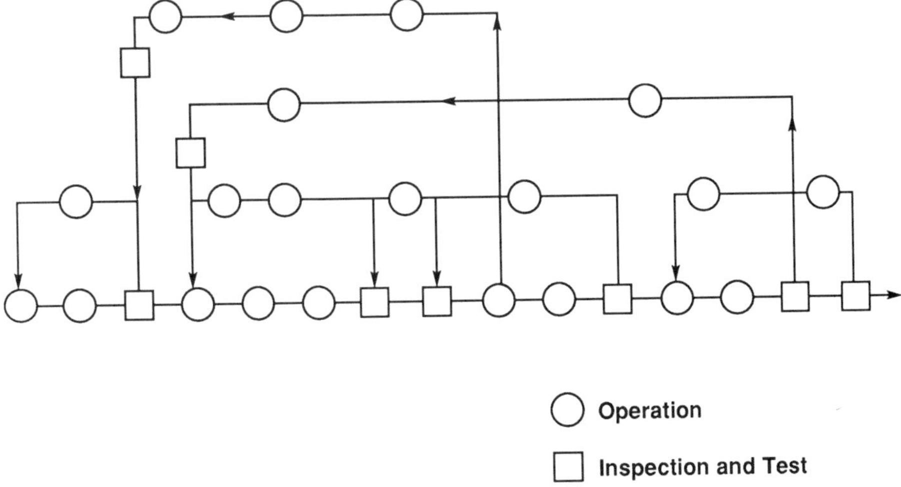

Figure 5.11 There are many times that rework exceeds the initial rejection.

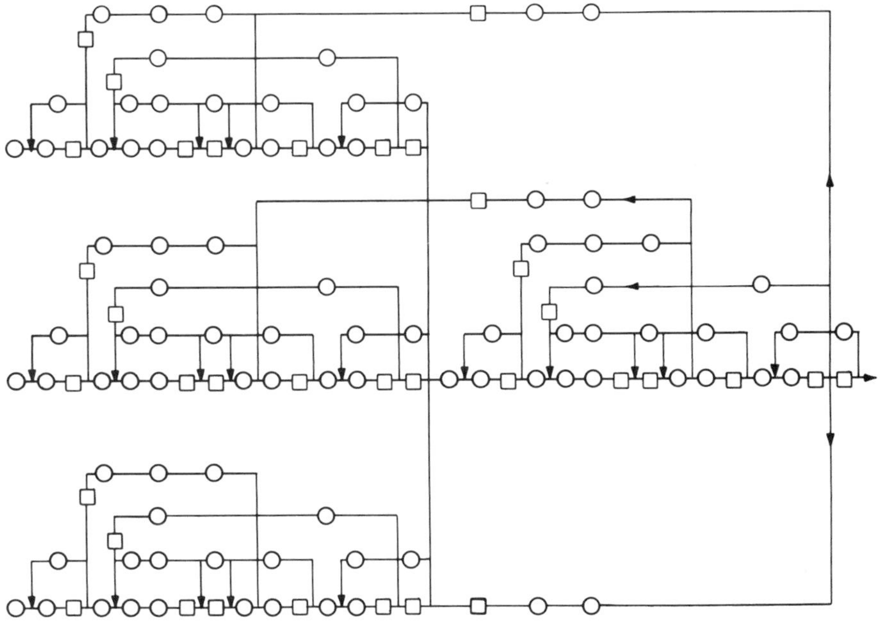

Figure 5.12 Combining three flows into an assembly creates a complex flow.

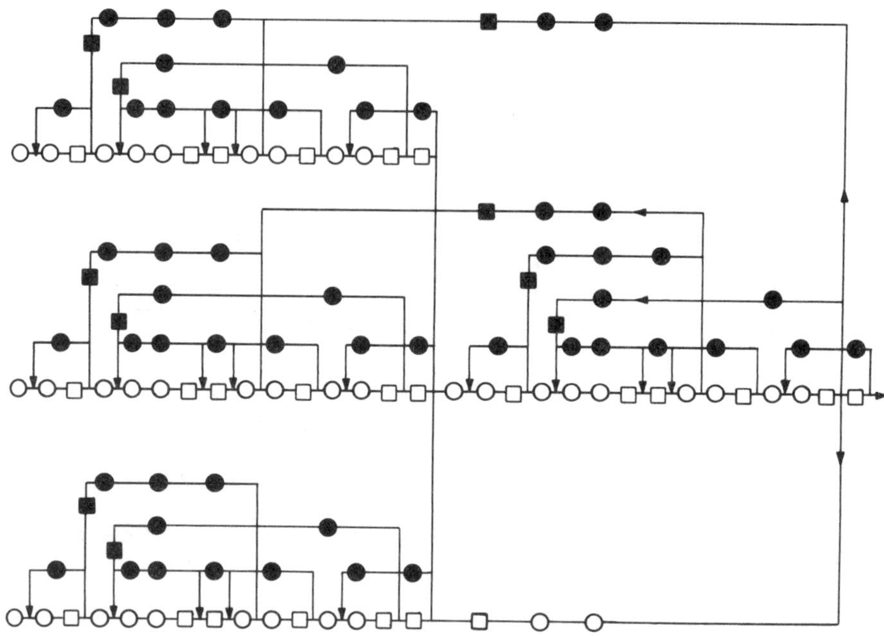

Figure 5.13 Highlighting the added rework and reinspection begins to present a picture of the potential magnitude of the problem.

of the rework. To provide readers with the visual perspective of the magnitude of this problem, Fig. 5.13 shows the added and extraneous operations.

Properly applied quality control concepts should result in less nonconformance and can permit less inspection and test. Figure 5.14 illustrates this type of flowchart with fewer inspection and test points. The

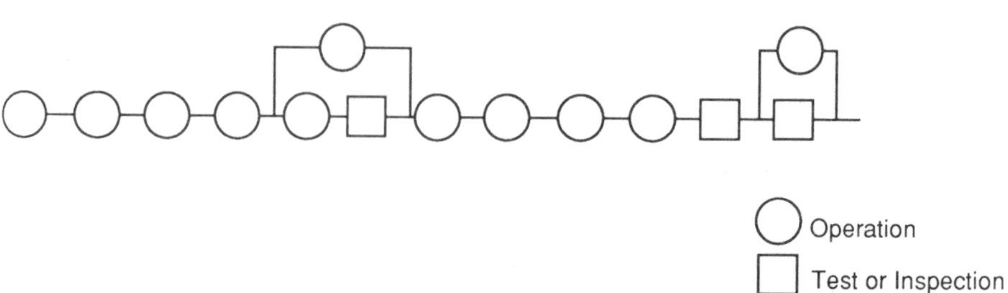

Figure 5.14 Illustration of the same number of operations with reduced inspection and test because yields are better in each step.

entire control system is dynamic and should change to meet product demands (whether the product is a service or physical material). Constant pressure for improvement must be applied.

Conclusions

The techniques discussed have been in effect in some companies since the late 1950s and early 1960s. More and more firms are realizing that newer methods must be introduced to improve products and meet competitive market standards and customers' rising expectations. Longer warranties in the future offer a major opportunity for advertising and sales promotion. Competitors are continuing the thrust toward improved product quality and reliability. The Japanese have demonstrated that product quality will capture and retain market share. A good product reputation rapidly spreads to other company products. To be competitive in the world marketplace, products must perform as advertised and must continue to perform with as few service calls as possible. A good quality cost system could help. See Table 5.4 for some questions to help evaluate a system.

In the manufacturing and service area, one way to meet the challenge is to develop a quality cost system. This will identify areas needing attention due to excess costs. Quality data analysis, discussed in later chapters, will indicate how problems create these high costs. Corrective measures can then be introduced to improve quality and reduce costs, with a resultant rise in profit, improve customer satisfaction, increased market share, and better competitive position.

Appendix A: Quality Cost Element Definitions

Internal Failure Costs

Rework Costs: Costs associated with the labor resulting when an item is reworked to acceptable condition after it has been rejected. In addition to actual rework time, these costs may include preparation of procedures for accomplishing the rework, establishing the control procedures, performing additional reinspections, and material required. Rework due to engineering change notices (ECNs) or customer-directed changes should be compiled separately.

Retest Costs: Costs associated with all retest resulting after initial failure of product in test. This includes all labor categories associated with retesting the equipment until it reaches the point at which it was

Table 5.4 Is Your Quality Cost System Effective?

1. Does your accounting system provide cost data that show prevention, internal failure costs, external costs, and appraisal costs?
2. a. If yes, are the reports available within several days of month end?
   b. If not, what is required to get the data quickly?
3. Is a monthly report necessary? Is it consistent with the other accounting (financial) reports?
4. Do the data provide visibility? Can we tell the value of each cost element? Can trends be detected? Are the data plotted, or in masses of computer reports that take too long to extract?
5. Are the cost elements given in absolute terms as well as in ratios to sales or to another base that provides an understanding?
6. Are data provided by department, plant, and product so that needed action can be determined?
7. Are failure costs too high? What action is planned? Does the action address the problem?
8. Are appraisal costs too high? What action is planned? Does the action address the problem?
9. Are rework costs available? How are they obtained? Are they accurate enough to base an intelligent decision on them?
10. Do the scrap costs include labor as well as material?
11. Do you know the cost of incoming, in-process, and final inspection and testing? Do the costs include all inspection, regardless of whether the inspectors and testers are in a quality control, manufacturing, or some other activity?
12. Are the expenses of appraisals justified by the internal and external failure costs; that is, do you have extremely low failure costs and high inspection costs? If so, can they be reduced? Are both costs high? What can be done?

rejected initially and becomes acceptable. Retest due to ECNs that are required to meet new requirements or retest required as a result of an ECN not specifically required due to product failure should be compiled separately.

Scrap Costs: Costs for material and labor associated with the fabrication of an item that is subsequently scrapped due to nonconformance to specification. In this category, all scrap should be counted, including that which is discarded by the manufacturing department (prior to reaching quality control) due to nonconformance. Fallout in manufacturing due to poor quality must be classified in this category. Scrap associated with cuttings, and chips are not included. The additional scrap resulting from an engineering change notice should be compiled separately because it provides additional insight into the source of the problem.

Troubleshooting Costs: Costs associated with the time required to isolate a problem identified in a test but which cannot be repaired until the specific cause is outlined.

Failure Analysis Costs: Costs associated with performing the analysis and preparing any associated reports.

Extra Operations Costs: Costs associated with adding an operation such as touch-up or trimming because the preceding operation did not perform the operation adequately.

Corrective Action Costs: Costs associated with performing corrective action as a result of inspection, audit, or design review action items. In addition to corrections on hardware, corrections on documentation and procedures are included.

Excess Inventory Costs: Costs associated with extra materials required to enable operations to continue. These include the cost of extra material purchased to allow for scrap in-line and extra material to support in-process rejections.

External Failure Costs

Warranty Costs: Costs to cover material, labor, and travel required to uphold warranty obligations to customers.

Customer Complaint Costs: Costs for travel and subsistence for persons responding to a specific customer complaint, as well as time spent by marketing, engineering, and quality assurance personnel to satisfy customers. Frequently, a trip is made partially as a sales venture and partially to respond to customer complaints. The portion associated with customer complaints should be included.

Customer Return Costs: Costs associated with receiving, evaluating, and repairing products returned by customers. Also included are shipping charges associated with return and reshipment of nonconforming products.

Added Service Costs: Costs associated with service operations such as appliance repair which must be redone because of inadequacies in the initial servicing operation. Includes all costs associated with product recalls.

Seconds Costs: Cost differential between prices for first-run quality and those charged for seconds.

Appraisal Costs

Incoming Inspection Costs: Costs for inspecting purchased material upon receipt. This should be a target for significant reduction or elimination.

In-Process Inspection Costs: Costs associated with performing inspections by quality assurance or manufacturing inspectors as material flows through the production line. It includes time spent in inspecting processes as well as in inspecting product.

Final Inspection Costs: Costs associated with performing final inspection.

Incoming Testing Costs: Costs of tests performed on all items purchased to determine acceptability. This should also be a target for significant reduction or elimination.

In-Process Test Costs: Costs associated with tests on items as they progress in the building cycle. Includes test monitoring and all costs associated with environmental testing.

Final Test Costs: Costs associated with the final performance and environmental testing on the deliverable end item.

Quality Audits Costs: Costs associated with performing audits on products, processes, and systems. However, customer audits and supplier audits should be a subdivision of this cost.

Calibration Costs: Costs associated with recall, calibration, repair, and redelivery of equipment or tools and gauges being calibrated.

Prevention Costs

Quality Planning Costs: Costs related to preparing a quality program plan to define all actions necessary to ensure that the quality and reliability requirements and standards will be met. This is generally the time for developing the program plan and for updating the plan.

Training and Motivation Costs: Costs related to preparing and implementing training programs, including the expenses of the instructor, the time of the participants, and efforts associated with motivation.

Process Planning Costs: Costs associated with developing and establishing process controls required by design. Include not only the quality assurance engineer's time but also the process or development engineer's time to set up the program.

Vendor Selection Costs: Costs for the time required by a vendor survey team to select vendors who are most likely to provide trouble-free equipment, materials, parts, and services.

Design Review Costs: Costs for the time the design review team and all other participants spend reviewing a design. Includes preparation time required for the formal design review.

Parts Selection Costs: Costs for the time of design and reliability engineers to select and specify reliable parts for a program.

Qualification Costs: Costs for the proving of the design through a testing program intended to expose products to more severe environments than would normally be encountered to provide a design safety margin. May include tests to destruction to identify product weaknesses.

Reliability Analysis Costs: Costs associated with such reliability activities as failure mode and effects analysis, reliability apportionment, maintainability prediction, and other reliability elements not associated with failure analysis.

References

American Society for Quality Control (1971). *Quality Costs: What and How,* ASQM, Milwaukee, Wis.

Hagan, John T. (1986). *Principles of Quality Costs,* American Society for Quality Control, Milwaukee, Wis.

Juran, J. M. (1965). *Managerial Breakthrough,* McGraw-Hill, New York.

Juran, J. M., and F. M. Gryna, Jr. (1970). *Quality Planning and Analysis,* McGraw-Hill, New York.

6 • Communications and the Process

Introduction

The problem to address next is how to take effective action to make improvements. The major benefit of a quality cost system is that it provides visibility into operations so that areas needing change can be identified. The quality cost system, however, does not help to determine what specific action is needed. To decide on a particular course of action it is necessary to gather other information—information that will indicate whether a design must be changed; whether a machine must be overhauled, maintained, or replaced; whether a process must be adjusted; whether training is necessary; or just what detailed action is required. To make this decision, additional data are needed, data that must be collected from the process. The purpose of this chapter is to discuss communications from processes to people. It is intended to show how the process can be made to inform you of its intentions so that you can be in a position to control the process to make it do what is desired.

Process Fundamentals

Let us begin by defining a process as a sequence of operations or events whereby specific results are achieved (or desired). Under such a definition, an entire manufacturing operation may be considered a process, or just a segment of the manufacturing cycle. A complete or partial service may also fulfill the definition. Examples of what might be considered a process include the manufacture of nickel–cadmium batteries, the manufacture of printed circuit boards, the manufacture of assemblies, the manufacture of parts from a multiheaded machine such as an automatic screw machine or plastic parts from a mold, the function of receiving inspection, the process of magnetic check reading, the production of a magazine or newspaper, and the response to a service call and correction of the problem.

Some of the reasons for controlling the process are to:

1. Maximize the percentage of good products produced initially.
2. Improve efficiency of operations.
3. Keep costs to a minimum.
4. Prevent generation of unreliable products or services.
5. Maximize profits.
6. Reduce material needs.
7. Improve field performance.
8. Achieve better reception by consumers.
9. Improve sales as a result of improved customer satisfaction.
10. Develop esprit de corps among the work force associated with pride in product or service performance.

As the process is operating it is spewing out information. We must know:

1. Which information is important
2. How to collect it efficiently
3. How to display it so that its message is clear
4. How to respond effectively to the message
5. How to measure the results
6. How to keep the benefits in place

In setting up communications, we must be sensitive to the many process characteristics and the ways in which the information that the process is trying to communicate can be obtained. Possibly the most difficult task is to determine the significant process characteristics. This is where process or systems knowledge comes into play. Those people responsible for designing a product, establishing a process, or setting up a service must become involved in determining what characteristics need to be measured and how these measurements should be taken. If data are being taken, a convenient format for displaying results must be developed. If data are not being taken, a way to collect and present the data may have to be developed. One must be careful to make sure that just enough data are taken to represent the process. All too often, too many data are taken and the system becomes cumbersome and collapses. It is better to err on the side of too few data than too many data. Also, recognize the dynamics at play. Converting theory to practice is not easy. We do not always take all the key variables into account; or we may measure the wrong variable. There are subtle effects that may be due to operators, equipment, materials, plant environment, and other areas that are not recognized. The way in which materials or processes change during a manufacturing operation may not be known precisely. Continued monitoring will indicate whether the proper characteristics and the proper limits have been instituted. Properly done, the communication (measurement) system can identify these needs.

A second factor necessary in communicating is to be sensitive about how a process may shift. Consideration should be given to whether process changes are long or short term, gradual or sharp, shift in level or shift in spread, or are cyclical or random. This evaluation helps to determine the type of information collection method to be used and the frequency of data collection. The nature of this variation may indicate the source of the problem provided that the communication system is set up properly.

A third factor needing evaluation is the type of information necessary. For example, should variables (measurements) be used or are attributes (go–no go) adequate? Variables data may be more time consuming to collect than attributes data but if displayed correctly can usually provide more information about a process.

A very major consideration is how the information should be collected. Are manual means adequate or should computers be used either by batch processing or on a real-time basis? Care must be exercised in making this decision because it is easy to fall into the trap of putting data into computers without first considering the cost of the entire data collection system and the potential cost savings as a result of using records for achieving control. Computer programs are great for collecting large masses of data and analyzing the results. The report formats, both input and output, need to be planned to provide simplicity and maintain output whose meaning is easily grasped. It is a good idea to try manual means or a personal computer before going directly into a massive computer data collection scheme. It is also desirable to consider data base systems to allow for other uses for information placed into the computer. Production status can be obtained from essentially the same data base as quality status, for example.

Keep in mind that the technology is advancing at breakneck speed, so that data collection methods are also changing rapidly. Bar-code scanning and optical character recognition are two methods that might facilitate data collection. Displays provided directly from measuring and test equipment can give excellent visibility. However, do not discount manual means of collecting and analyzing process data. When a process is complex, or when it is necessary to merge data from multiple sources to get a picture of what is going on, the computer offers an advantage. Also, when automated equipment is under computer control, the computer can offer a significant advantage by providing direct feedback to adjust processing parameters.

Case History 1: Manufacturing Printed Circuit Boards

An example should help illustrate some of the points. In a printed circuit board manufacturing operation, there are many liquids in solution, each with several variables that need to be controlled. Many chemical processes are controllable by automated means, in which case deviations from prescribed performance can be detected and error signals fed back to servo controls to adjust solution temperatures, pressure, concentrations, pH, contamination levels, or other parameters. The problem in this situation is in knowing the allowable variation and the correct levels of these parameters. In other instances, measurements of solution parameters must be made manually. The traditional method of doing this is to prepare forms showing the solution parameters and acceptance limits. A record is maintained of the results of periodic analyses by filling in the value for the solution characteristic in a blank space provided—see Table 6.1. The record is provided to the shop floor supervisor, who is responsible for seeing to it that the tank containing the solution is adjusted to bring its value within specification limits. In an ideal world the supervisor does this promptly after receiving the data from the chemist and the solution is corrected. In the real world, the supervisor intends to have the solution adjusted promptly but before he has a chance to do so is interrupted by several other crises and does not get around to having the solution fixed until hours or days later, if at all. Even if we assume that the solution adjustments are made and a recheck is taken, this is still not adequate to compete in a world-class marketplace.

Table 6.1 Results from Solution Analysis of Plating Tanks[a]

PRINTED CIRCUIT BOARD SOLUTION CONTROL RECORD
Table A: Copper Analysis

	Tank 1	*Tank 2*	*Tank 3*	*Tank 4*	*Tank 5*	*Tank 6*	*Auto A*	*Auto B*	*Auto C*	*Auto D*	*Percent out of limits*
Copper	1.07	1.12	1.17	1.09	1.07	1.23	1.20	1.40	0.98	0.99	
Pyro	0.07	0.05	0.06	0.07	0.07	0.06	0.08	0.08	—	—	
Ammonia	10.1	11.2	11.1	10.9	10.7	10.8	17.3	10.8	10.7	10.4	
Pyro/Cu ratio	1.1	1.2	1.1	1.1	1.4	1.5	1.2	1.1	1.1	1.1	
Specific gravity	0.92	0.97	0.93	0.94	0.95	0.92	0.91	0.96	—	—	
pH	9.2	9.5	9.7	9.8	9.2	9.8	9.7	9.8	9.7	9.8	
Temperature (°F)	110*	115*	112*	113*	114*	114*	115*	112*	112*	117*	
Parameter/total											
Total %											

Table B: Nickel Analysis

	Auto plate	*Tank 1*	*Percent out of limits*
Nickel sulfate	10.2	11.4	—
Nickel cloride	10.7	11.2	
Boric acid	8.7	8.7	
pH	9.0	9.1	
Parameter/total			
Total %			

(continued)

Table 6.1 *(continued)*

Table C: Gold Analysis

	Tank M-1	*Tank M-2*	*Tank U-1*	*Tank U-2*	*Tank auto*	*Tank strike*	*Percent out of limits*
Gold	1.1	1.4	1.7	1.8	1.1	1.3	
Cobolt	0.12	0.14	0.14	0.13	0.14	0.10	
Baumé value (deg)	15	15	15	12	10	15	
pH	9.7	9.8	9.6	9.5	9.3	9.7	
Parameter/total							
Total %							

Table D: Tin–Lead Analysis

	Auto high throw	Manu high throw	Percent out of limits
Stannous tin	—	—	—
Lead	—	—	—
Fluoboric acid	—	—	—
Specific gravity	—	—	—
Parameter/total	—	—	—
Total %			

[a] *, Solution within standard limits; X, solution out of control; —solution not analyzed during typical frequency.

Why not? Because the record is placed on file for possible future use, which presents the first problem. There is a loss of continuity in the process control because the back records are seldom reexamined to determine trends. Only when serious difficulty is encountered do the past records get reviewed. Then it can become time consuming—and all the while, lower yields or inferior quality are being experienced. Even when computers are used, tabulated data that are provided are generally inadequate to provide good process communications because it is difficult to spot behavior patterns quickly. Second, corrective action takes too long. We can assume that as the value of the solution parameter varies more widely from its nominal value, the resultant product is not as good as product that would be produced if the solution were at a constant nominal value. Thus more variability will exist in the product. In the third place, nothing is being done to reduce the amount of analysis being performed so test costs will continue. In the quality cost vernacular, appraisal costs will be level and failure costs (internal and external) will be higher. There is no self-correcting mechanism to reduce these costs.

Does this mean that there must be a substantial capital investment to improve this situation? Perhaps yes and no. What we need is a motion picture of the process, not a snapshot. Suppose that we use a simple graph to plot the solution parameters. In an actual situation where this was happening, everyone said that a graph was a waste of time. But an aggressive quality engineer took the data from the files and plotted them in a series of graphs. Some of them showed good solution stability, but about six solution characteristics exhibited the typical performance shown in Fig. 6.1, region a. Each time the value went out of control an adjustment was made to bring it within limits. The adjustment, though, was never enough to bring the solution characteristic to its nominal value. As a matter of fact, most solution values averaged out to the upper specification limit. This was never visible from the weekly solution checks. Everyone knew that there were some solutions requiring more adjustment than others, but they simply assumed that it was just the nature of the process.

Even after the graphs displayed the problem, there was much discussion about how the adjustment should be made—and whether it should be made at all. After all, the process had been behaving this way for years. Maybe the specification limits should be changed instead. The electrochemists prevailed and the solution was adjusted to a level below the specification average (see Fig. 6.1, region b). The first result was that the solution that exhibited this behavior required less adjustment because the process changed only by increasing in value. By adjusting the solution to a lower value it could remain within specification limits longer. This was done, but the results were poor, simply because the solution value was allowed to vary throughout the full range of the specification and the product characteristics have even greater variability. Incidentally, the end results took several months to become apparent.

Fortunately, the quality engineer had anticipated this problem and requested that the amount of additive used to adjust the solution value be recorded on the chart. Collectively, the group, consisting of the electrochemist, the supervisor, and the quality engineer, decided that they

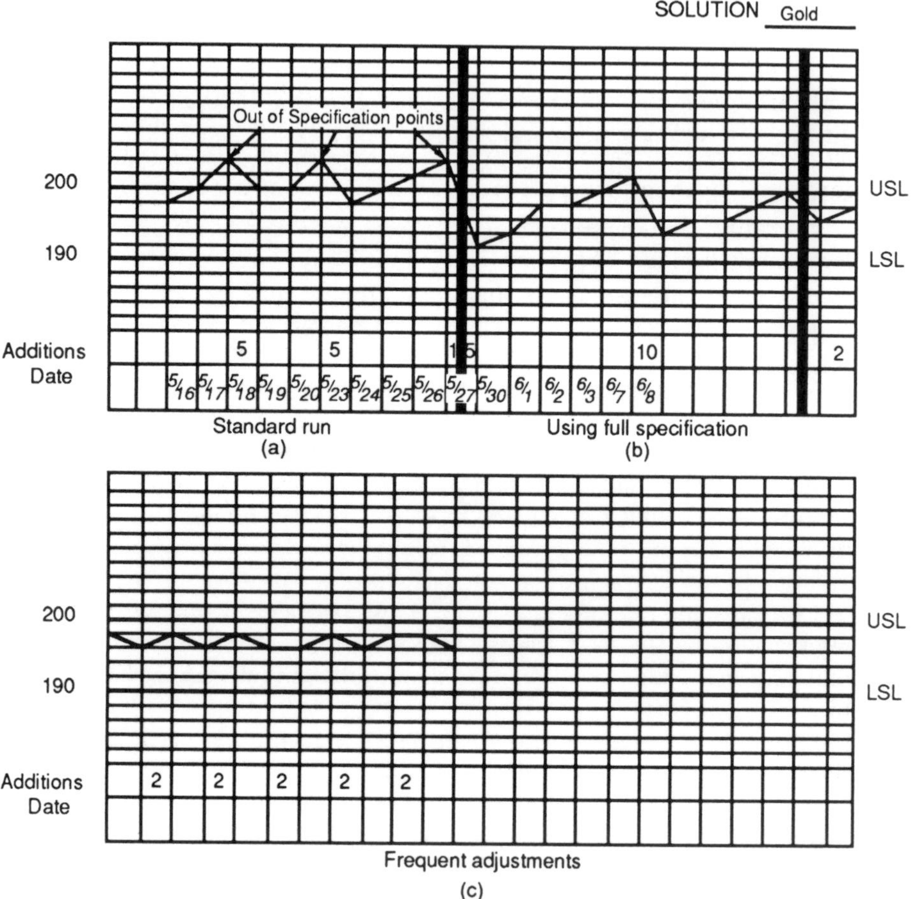

Figure 6.1 Graphs of solution control tabulated values. Table 6.1 provides information for one point on the graph.

could simply take the weekly quantity that was added, determine the daily amount, and adjust the solution each day. The results were spectacular. The solution could be maintained much closer than the specification required and the product became very uniform—outperforming all competition (see Fig. 6.1, region c). As the group gained confidence in the process, the solution testing was reduced by 75%, thereby freeing up one analyst. This reduced both appraisal costs and failure costs. This is what is necessary to compete in a world marketplace: lower costs and better product quality.

Administrative Data

An important form of quality data is administrative data. These data are intended to guide action by informing the reader of a general situation rather than identifying a specific problem. Such data are usually more management oriented and are intended to show broad-based performance—such as how a division or product line is performing. Here, too, we must be conscious of how data are displayed so that they convey intelligence. The essential features are:

1. *Information content:* what should be collected and displayed
2. *Accuracy:* vital to make sure that all the information is believed
3. *Clarity:* to present information concisely that displays what you want it to display
4. *Brevity:* to provide the reader with salient points
5. *Scannability:* to enable the reader to grasp quickly what is happening and to enable decisions to be made quickly
6. *Timeliness:* so that data reflect what is happening currently

One aspect particularly lacking in many reports is *scannability.* The report must be designed for the busy manager or engineer who has other reports and information to digest. By making the report scannable, the significant elements can be absorbed quickly and rapidly incorporated into a plan of action. Otherwise, the report is filed for future reference and nothing will get done!

An example of a computer-generated administrative report is shown in Fig. 6.2. All the information is there but I defy you to interpret it. These same data are shown in Fig. 6.3 in a format that can readily be grasped. Although it is generally desirable to keep the number of pages to a minimum, in this instance, two pages are used to display the data instead of one page. But multiple graphs on a page is a strategy that enables comparisons among operations as well as determining trends within a single operation. In displaying data such as these, an added feature is to use the same scale (sacrificing some precision) to enable relative performance to be observed.

When using a computer for data collection, if data cannot be input directly or if a terminal is not available, an input form has to be completed. This should be designed to require a minimum of effort on the part of the user—preferably, check marks should be used, not written material. When possible, bar coding, optical character recognition, or automated computer input methods should be used. New automated inspection and test systems can generate data in many formats to make

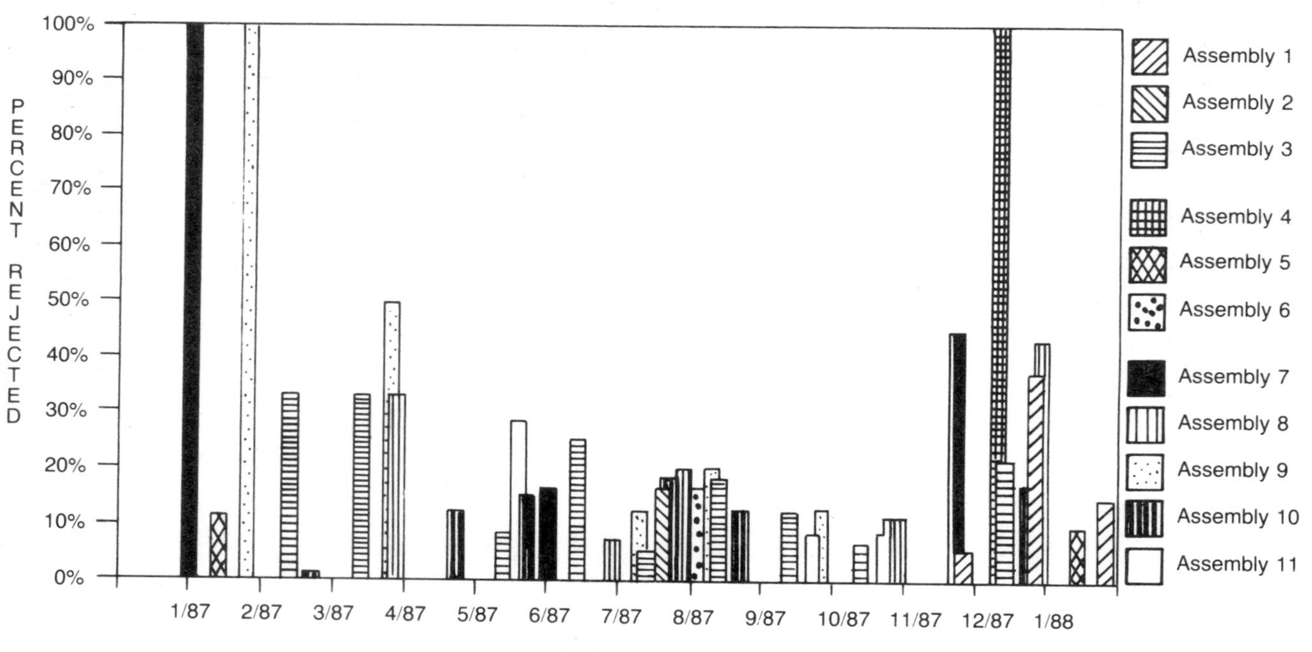

Figure 6.2 Computer generated report combining departments to obscure intelligence without meaning to do so.

analysis easier or self-evident and indicate an action. These machines should be considered an essential information source and not used merely for a required inspection or testing operation. Do not forget: The main purpose of testing and inspection is information gathering to take quick action, not determination of product acceptability.

Sources of Variation

As stated earlier, thought must be given to sources of variation so that records will reflect true performance. In every possible situation, the reporting system should require little or no auxiliary analysis. Instead, the analysis should occur automatically to minimize delays that make action too late to accomplish something. Some sources of variation that have to be explored may be:

1. Different machines
2. Different heads on multiple-head machines—or different cavities from a mold
3. Different assembly lines
4. Differences among operators or inspectors
5. Differences among suppliers or differences with time from a single source
6. Depletion or enhancement of materials
7. Environmental changes
8. Methods differences
9. Inspection or test differences
10. Different plants
11. Voltage or power changes
12. Differences in design
13. Shifts in process variables with respect to time
14. Other sources

The records should provide insight into differences in performance among these factors. Let's consider an example. Suppose that there are machines producing a product that is 10% defective. If you do not know the performance of each machine individually, you are operating blind. If each machine is producing at a rate of 10% defective, you have one bit of intelligence that may indicate some form of corrective action. On the other hand, if one machine is producing product's at a 30% defective rate and the other two machines at zero percent defective, something better be done—and quickly. More often than not,

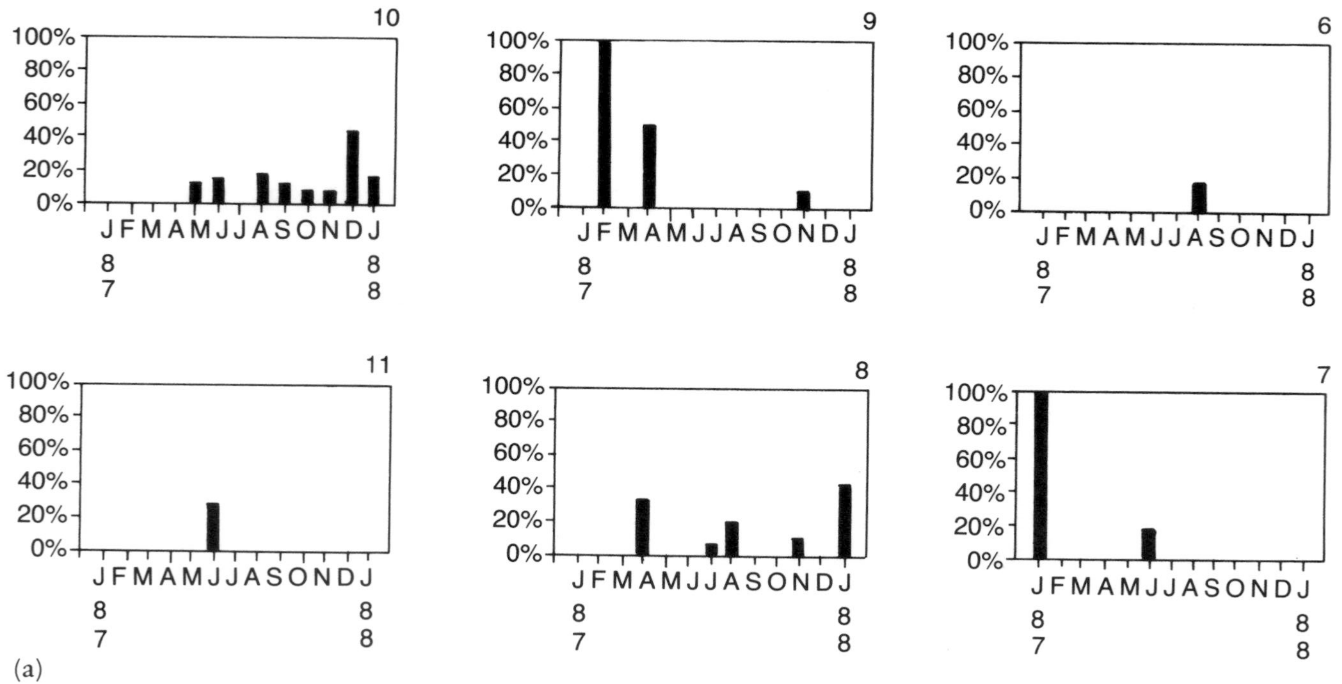

(a)

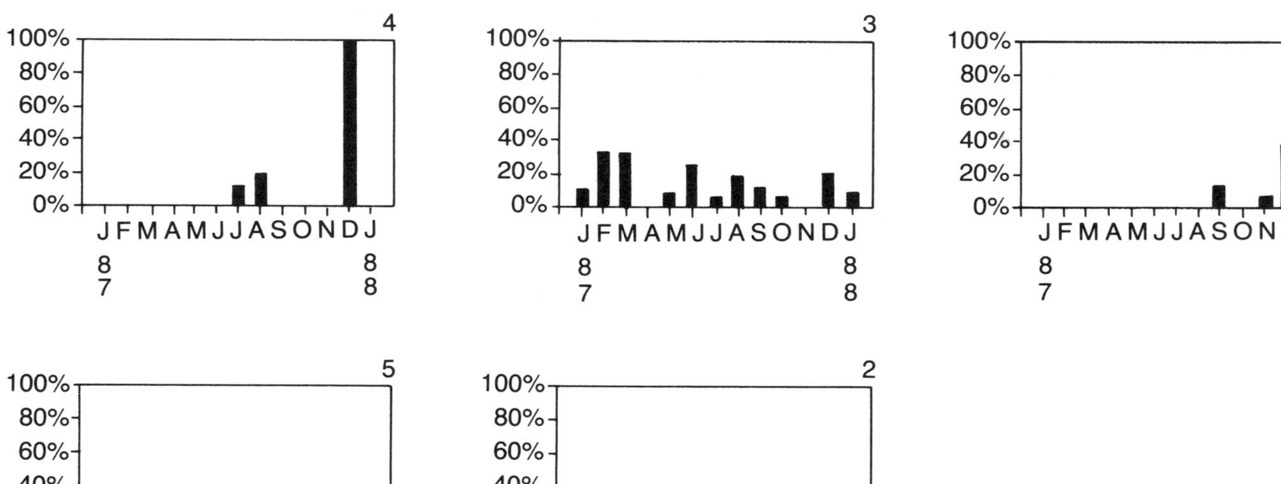

Figure 6.3 (a) Data from chart in Fig. 6.2 separated into individual assemblies, providing more visibility into operations. (b) Remaining data from Fig. 6.2. Although two pages are required instead of one, visibility is much better.

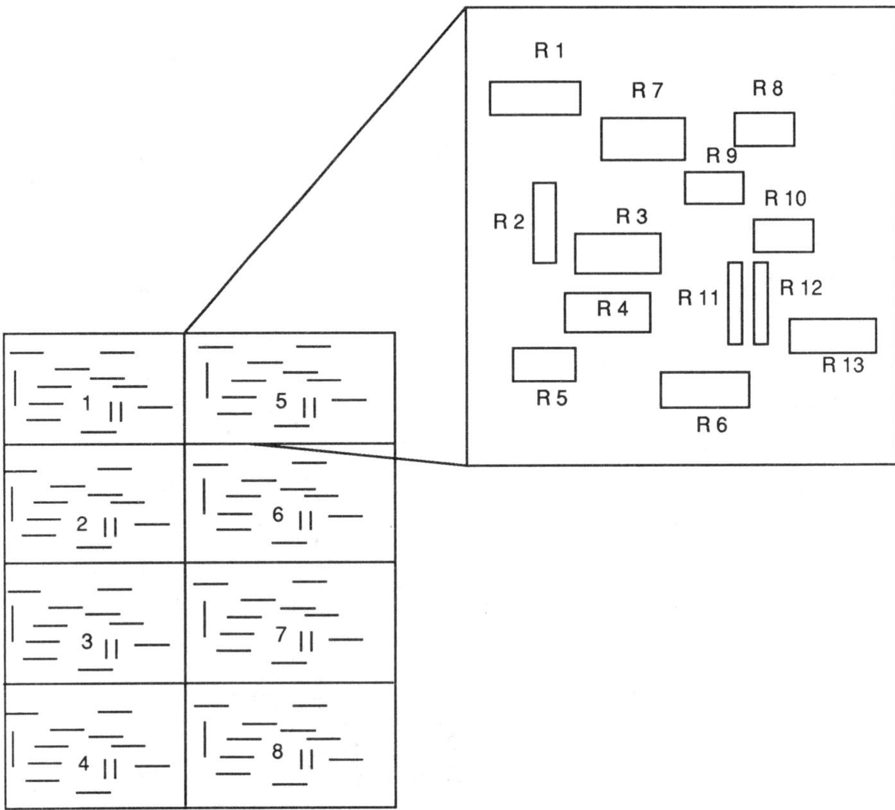

Figure 6.4 Substrate layout showing eight patterns that are screened at one time. Each pattern has 13 resistors, as shown on the blowup.

there are differences among machines—or among heads in a multi-headed machines—that are going undetected. These differences must be made visible so that corrections can be made. Continued pressure must be exerted to obtain fewer defects. In preparing to communicate with a process, these items must be taken into account. Before other aspects are discussed, let's consider another example.

Case History 2: Data Gathering in Printed Circuit Board Manufacture

In manufacturing thick-film hybrid circuits, resistors are screened in an eight-up pattern (eight patterns screened on one ceramic at a time). Thirteen resistors were contained on one pattern (see Fig. 6.4). Since resistors cannot be screened to the exact value, they must be trimmed to value. The trimming operation can only increase the resistance since it requires removal of material. Resistance readings were taken on the initial pieces to determine whether the parts could subsequently be trimmed to the proper value. The record shown in Fig. 6.5 was prepared by an operator using a resistance bridge. An engineer examined these data and used process knowledge to decide whether the lot could be adjusted to within specification limits.

A variation on the record keeping was introduced involving the use of a frequency distribution. These same data were plotted in frequency-distribution form (see Fig. 6.6). Frequency distributions display the frequency of occurrence of a measurement versus the value of that

CIRCUIT TYPE

TITLE: DATE: ______

		R1	R2	R3	R4	R5	R6	R7	R8	R9	R10	R11	R12	R13
#1	1	.973	712.11	37.9	15.8	41.0	5.72	37.6	519.0	549.9	64.0	35.8	1.25	11.6
	2	.914	753.4	30.4	12.2	34.9	5.77	34.5	458.9	550.8	67.1	35.3	1.03	12.9
	3	.871	876.4	36.1	11.5	31.3	3.71	31.4	356.6	692.5	62.5	33.7	1.07	10.6
	4	.815	754.1	26.5	8.8	28.8	4.70	25.6	339.6	883.5	63.1	29.5	.96	8.9
	5	1.09	710.5	31.9	13.7	40.6	6.73	37.8	418.3	700.82	68.1	35.8	1.10	10.3
	6	.998	843.7	32.1	15.1	37.5	6.54	32.2	383.1	769.5	61.5	29.7	.97	9.5
	7	.942	608.4	25.3	14.1	32.5	5.15	27.8	343.3	655.3	54.9	26.5	.96	8.6
	8	.961	719.9	25.1	+8.5	34.5	4.60	25.4	417.7	743.2	60.6	25.5	.87	7.9
	AVERAGE	.979	747.3	30.7	12.4	86.1	5.36	31.53	354.6	655.7	62.7	31.5	1.05	10.0
#2	1	.567	613.8	37.1	15.1	37.5	5.13	37.5	386.4	672.7	57.1	32.1	.86	10.2
	2	.519	553.4	38.1	15.7	34.2	4.50	31.7	343.0	655.1	57.0	30.4	.77	11.6
	3	.501	559.3	24.7	8.7	30.8	5.12	32.7	239.0	702.1	68.5	29.7	.42	10.3
	4	.638	546.8	27.7	8.1	28.8	4.42	30.7	276.0	746.1	63.6	28.6	.83	9.1
	5	.721	746.1	31.3	11.4	38.9	7.01	33.6	373.3	643.3	67.5	37.2	1.00	10.7
	6	.653	585.1	29.7	14.2	37.7	5.15	32.6	314.8	718.2	60.3	25.7	.85	9.9
	7	.512	704.1	26.9	17.9	31.3	4.19	25.6	251.9	827.7	53.2	27.3	.81	8.3
	8	.537	535.2	27.3	9.5	32.4	3.91	25.0	248.9	643.1	56.3	27.1	.81	8.1
	AVERAGE	.581	596.5	30.4	12.6	34.0	4.93	31.2	304.1	701.0	60.4	29.8	.87	9.8
#3	1	.869	533.1	34.6	13.9	35.4	5.44	32.4	450.6	751.5	53.1	34.6	1.06	11.2
	2	.728	583.8	28.3	10.6	30.9	5.30	29.9	406.3	765.5	51.6	32.2	.98	11.4
	3	.719	542.5	24.7	9.8	30.7	4.78	30.2	299.3	744.1	58.8	30.7	.83	9.6
	4	.740	678.7	23.6	7.7	25.5	4.01	26.6	275.6	754.3	56.6	28.9	.91	7.2
	5	.831	702.6	31.0	13.5	38.9	5.15	32.6	349.4	851.2	67.6	33.2	.94	7.7
	6	.752	716.1	32.7	13.4	37.4	5.05	29.5	317.9	815.2	60.8	21.5	.85	7.3
	7	.839	857.3	35.8	12.9	33.5	4.35	24.5	301.7	717.8	55.3	24.5	.75	7.8
	8	.792	744.9	29.0	9.2	31.1	4.02	22.1	363.9	838.7	53.4	24.1	.79	7.7
	AVERAGE	.784	670.0	30.0	11.4	32.9	4.76	28.5	345.6	886.5	57.1	28.7	.90	8.7
#4	1	.745	637.1	34.7	15.2	38.6	4.57	31.2	516.9	908.1	60.1	33.7	1.01	11.2
	2	.515	647.6	39.6	12.1	34.5	4.74	33.5	475.4	836.6	62.1	31.6	.98	12.2
	3	.553	665.6	25.8	10.9	32.5	5.18	32.9	373.7	853.5	61.8	33.1	1.00	10.3
	4	.713	723.3	27.6	9.5	27.8	4.01	27.8	277.5	7.00.0	56.4	30.5	.94	7.9
	5	1.005	701.3	27.1	11.3	44.8	3.51	35.7	451.1	1014.0	68.7	31.9	.99	11.3
	6	.515	755.3	30.1	14.0	36.3	5.25	31.2	383.9	910.0	61.1	29.6	.82	11.2
	7	.917	595.2	28.9	13.7	35.6	4.60	25.7	373.6	810.9	55.3	28.7	.82	9.1
	8	.819	630.3	23.3	9.8	29.9	4.22	25.1	399.4	900.0	60.9	21.9	.83	8.8
	AVERAGE	.697	669.5	29.6	12.1	35.0	4.51	30.4	406.4	866.6	60.8	30.1	.92	10.2

Figure 6.5 Tabulated values of resistance for 13 resistors on each of eight patterns. There are four samples shown.

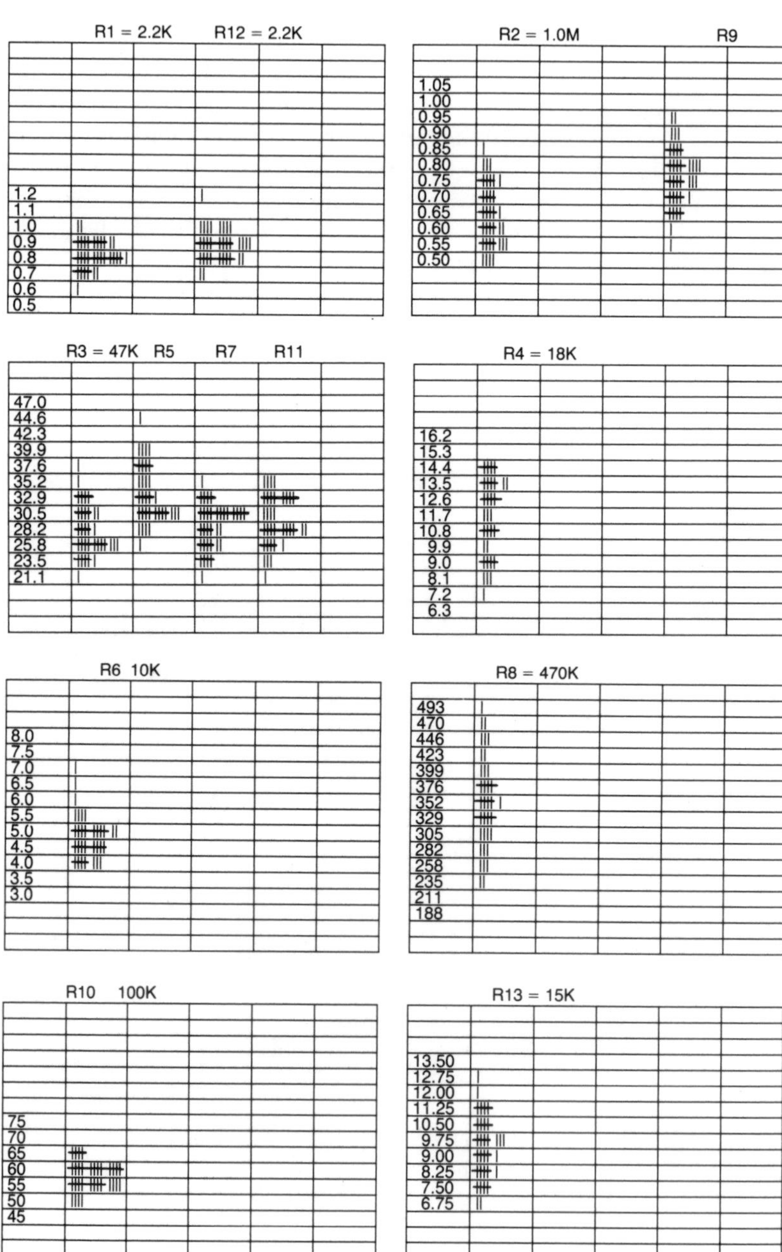
R1 = 2.2K
R12 = 2.2K
1.2
1.1
1.0
0.9
0.8
0.7
0.6
0.5
R2 = 1.0M
R9
1.05
1.00
0.95
0.90
0.85
0.80
0.75
0.70
0.65
0.60
0.55
0.50
R3 = 47K
R5
R7
R11
47.0
44.6
42.3
39.9
37.6
35.2
32.9
30.5
28.2
25.8
23.5
21.1
R4 = 18K
16.2
15.3
14.4
13.5
12.6
11.7
10.8
9.9
9.0
8.1
7.2
6.3
R6 10K
8.0
7.5
7.0
6.5
6.0
5.5
5.0
4.5
4.0
3.5
3.0
R8 = 470K
493
470
446
423
399
376
352
329
305
282
258
235
211
188
R10 100K
75
70
65
60
55
50
45
R13 = 15K
13.50
12.75
12.00
11.25
10.50
9.75
9.00
8.25
7.50
6.75

measurement. This enabled a number of characteristics of the process to be determined:

1. The level at which the process is running or its average ($\overline{X}$).
2. The spread of the process (standard deviation or sigma, σ).
3. Whether or not the process is multimodal (operating from a multiple set of chance variables).
4. By selecting the cell intervals as 5% of design value in this instance, the relative level of each resistance value in relation to the others can be determined.

Scanning these frequency distributions disclosed the following:

1. R1 and R12 are too low and should be adjusted upward.
2. R5 and R11 have tails on the high side. Most of the high values came from position 1 of the eight-position substrate. Further analysis showed that position 5 also had high values. The substrate was numbered as shown in Fig. 6.6, showing high values at one end of the substrates. The engineer recognized the effect of a tilted platen used in the screening process that allowed the thickness of the resistors in locations 1 and 5 to be different from the thickness in the other location, causing difference values.
3. R4 also has a bimodal (double-peak) distribution because the high values were from positions 1 and 5.
4. R8 has excessive variability, but no explanation was found.

This information could not possibly have been determined from the previous tabulation. The frequency distribution plots identified a tilted platen in the screening operation. It was corrected and higher yields and faster trim times resulted. The record-keeping format was revised accordingly, reducing the time the operator took for record maintenance. Subsequent changes using computerized data analysis were introduced, but the basic distribution analysis is still in use except that the distributions are plotted by the measuring instrument.

Case Histories 1 and 2 illustrate improved process communication in which a record-keeping system was revised to reduce record-keeping time and improve knowledge of what was occurring in the process.

Figure 6.6 Frequency distribution of resistance values showing bimodal values for R4 and R11 and long tails on R5,R8, and R3.

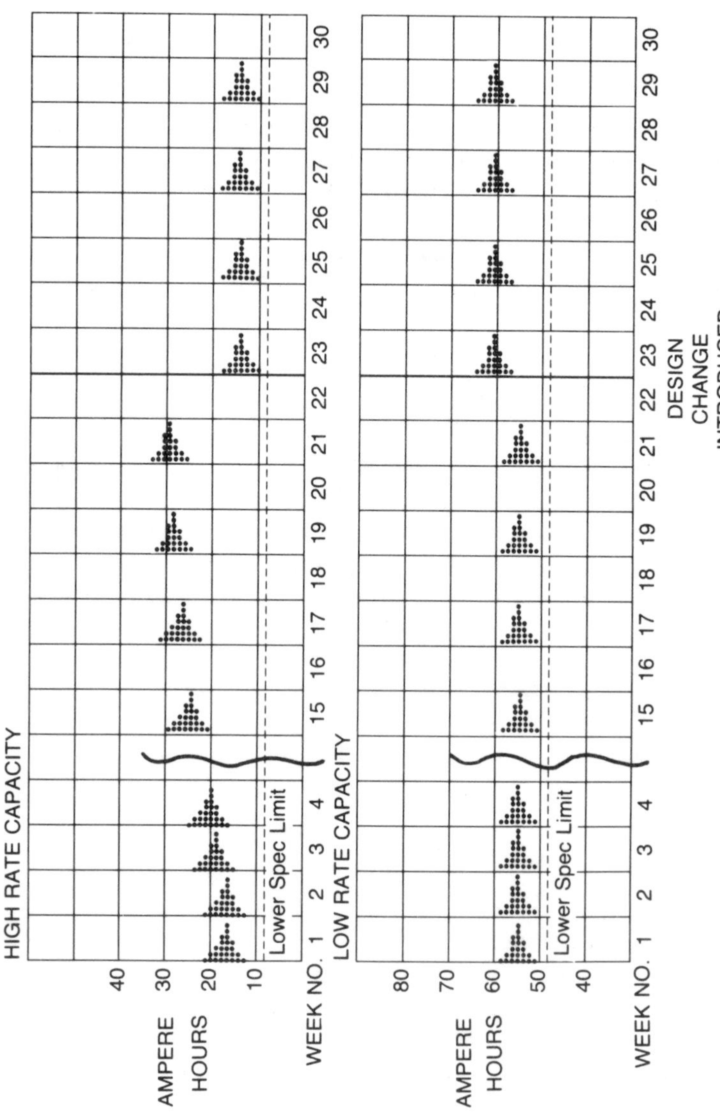

Figure 6.7 Battery discharge capacity.

Case History 3: Battery Capacity Records Example

In another, somewhat different situation, nickel–cadmium batteries were being manufactured. The customer required that the discharge capacity be provided at both high and low current rates. These data were shipped with each weekly lot. In most situations the tabulated data would be filed away and probably forgotten. This time, the clerk handling the paperwork was instructed to prepare a frequency distribution in dot format where each dot represented a measured value to provide continuing process visibility (see Fig. 6.7).

In this example, as process improvements were introduced in the plant, the battery capacity at the high current discharge rate gradually increased while the battery capacity at the low current discharge rate remained marginal. The process picture provided by this 6-month record indicated a possible design change that would raise the low current capacity and lower the high current capacity. This was accomplished by using fewer, thicker plates in the battery at a sizable cost reduction of 33%. The change was introduced and worked so well that it highlighted a new method of adjusting battery performance in all other product types. This opened the way for major cost savings throughout the business and led to an overall improved product. As a result of this approach, large new contracts were won in competition and a principal competitor was forced to seek shelter through acquisition by another company.

Using the Computer for Data Collection

Let's now consider the use of process information in a manufacturing operation where either inspection or test data are collected by computer. Automation has become a significant factor in electrical testing. This has led to the ability for rapid diagnostics by computer, with the further possibility of listing classes of defects by cause. This is done by presenting data in ordered format by equipment type, operator, department, foreman, inspector, or however requested information is effectively presented. This is also done by limiting data to the top offenders—in the format shown in Fig. 6.8 to the 15 most significant problems—information can be used directly to isolate problem cases. In addition, details are shown on each item (Fig. 6.9); reading from right to left, the defect classes are presented in sequence of most to

TYPE 1 REPORT DEFECT LEVELS MONTHLY TOTAL TEST–JUN 83 REPORT IDENTIFIERS: 006

LISTING OF: OPERATION REPORT DATE 07/05/83

DEFECTS RELATED TO: SOURCE: ALL ACTIVITY

PERIOD FROM 06/01/83

15 MOST SIGNIFICANT ENTRIES TO 06/30/83

					CURRENT DEFECTS			PAST		PERIODS	
CODE	NAME	ANNUAL COST	TOTAL UNITS	REJ UNITS	TOTAL DEF	% REJ	DEF/ UNIT	DEF/ 100 HR BUILD	% REJ	DEF/ UNIT	DEF/ 100 HR BUILD
	OVERALL		7190	2272	5222	31.5	.72				
A	ITEM 1		56	41	301	73.2	6.80				
B	2		254	51	160	20.0	.62				
C	3		91	70	157	76.9	1.72				
D	4		40	55	141	91.6	2.35				
E	5		40	31	131	77.5	3.27				
F	ITEM 6		46	2	120	4.3	2.60				
G	7		34	27	105	79.4	3.08				
H	8		32	27	103	84.3	3.21				
I	9		30	27	101	90.0	3.36				
J	10		18	15	100	83.3	5.35				
K	ITEM 11		194	79	96	40.7	.49				
L	12		60	45	95	75.0	1.58				
M	13		52	39	93	75.0	1.78				
N	14		69	46	76	66.6	1.10				
O	15		60	30	76	50.0	1.26				
	225 OTHER		6094	1687	3287	27.6	.53				

Figure 6.8 Computer-generated data showing top categories.

least significance. By scanning the upper left corner of data, the largest contributors to a problem are presented.

This type of format is well suited to use in a corrective action program. It is not well suited, however, as a report to higher management because too much detail is presented. A more appropriate management report is shown in Fig. 6.10. This report presents trends and enables a quick review to ascertain that the department is indeed operating effectively. By placing multiple products on a single page, interproduct comparisons are easier to observe, along with the normal evaluation for absolute level and trends. This is not the only type of management report, but it has proven to be an effective one. To obtain scannable results, the axes of the graphs must be the same.

Control Charts

No discussion of this type would be complete without a brief discussion of a statistical control chart application. A control chart for the average, $\overline{X}$, and range, R, is a very sensitive indicator to process shifts. It is the foundation of statistical process control (SPC) and is discussed in Chapter 7. Control limits can be determined as in Chapter 7 or by using any book on quality control (see Besterfield, 1986; Grant, 1980; Juran, 1988). Sample sizes are usually limited to three to six pieces selected periodically throughout production and $\overline{X}$ and R are calculated. Care must be exercised in determining this subgroup (here again, the reader is referred to existing texts). This technique is ideal for controlling a single characteristic in processes such as weight, weld strength and dimensions. A sample control chart is shown for integrated circuit wire bonding strength (Fig. 6.11).

Averages plotting out of control will occur 3 times out of 1000 by chance alone. Thus an out-of-control value is almost certain to be due to an assignable and therefore an eliminatible cause. Other indications of shift also exist; for example, out-of-control points on the range chart are indicative of a widening in process spread.

Case History 4: Data Gathering from a Molding Operation

A final example of process communication deals with a multiple-cavity mold (see Frey, et al., 1968). In this instance, toothpaste caps were be-

TYPE A REPORT COMPONENT DEFECT DISTRIBUTION — REPORT INDENTIFIERS: 006

REPORT DATE 07/05/83

DEFECTS RELATED TO: OPERATION XYZ — SOURCE: ALL ACTIVITY

15 MOST SIGNIFICANT ENTRIES — PERIOD FROM 06/01/83 TO 06/30/83

					CURRENT DEFECTS				PAST PERIODS		
CODE	NAME	ANNUAL COST	TOTAL UNITS	REJ UNITS	TOTAL DEF	% REJ	DEF/ UNIT	DEF/ 100 HR BUILD	% REJ	DEF/ UNIT	DEF/ 100 HR BUILD
A	ITEM 1		56	41	381	73.2	6.80				

COMPONENT			10 MOST SIGNIFICANT DEFECTS										
PART NUMBER	LOCATION/ NAME	TOTAL DEF	RES OTOL	RES WPRT	DIOO RVRS	XSTR SHRT	WIRE MISW	SOLD SHRT	CAP MISS	XSTR BXDN	CAP SHRT	SOLD MISS	OTHER
	OVERALL	381	195	40	22	14	11	7	7	6	6	5	68
	R–10	31	28	2									1
	R–12	27	22	5									
	R–14	17	14	3									
	Q–8	15					11						4
	Q–7	14						7				5	2
	R–2	13	10	2									1
	R–7	12	12										
	R–5	9	9										
	Q–2	9				5				2			2
	T–1	8											8
	R–22	8	8										
	R–14	8	4	4									
	C–72	8							7				1
	R–4	7	7										
	R–8	7	5	1									1
120 OTHER		188	76	23	22	9				4	6		48

Figure 6.9 Detailed data showing nonconformities by rank.

ing molded in a 48-cavity mold. Parts had a high reject rate due to short shots (incomplete molding). This reject rate continued unabated despite the efforts of engineers to resolve the problem. Capital expenditures for new molds were under active consideration until a clever quality engineer decided to classify the defects by mold mark and developed a very peculiar pattern (Fig. 6.12). When she then drew a picture of the mold cavity layout (Fig. 6.13), the defect pattern became obvious. This indicated that a change in mold gating was the solution to the problem. The change was made, defects fell to zero, and three inspectors were transferred to production operations. This is a good example of thinking like a process.

Other Considerations

Let's return now to several other aspects. Thought has to be given concerning for whom the report is intended. We have discussed administrative or management reports and operating reports to some extent. Reports must contain sufficient information to enable action to be taken by the people receiving a report. There are always those who like to receive copies of everything just to feel they are on top of things. It may be necessary to satisfy this desire. On the other hand, for a report to be worthwhile, it must be used. Continued evaluation must be applied and corrections instituted if the entire communications system is to be effective. Methods for obtaining corrective action are discussed in Chapter 8. Suffice it to say that the report must trigger action or provide information indicating that no action is required.

Design automation is being used increasingly. While feeding back data continues to be important in this approach, greater emphasis must be placed on incorporating criteria into the computer design rules so that initial designs are created that avoid mistakes of the past. This approach, coupled with simulation or computer-aided analysis, will also assure better designs. When manufacturing engineering is added in the design cycle, manufacturability is further enhanced. This concurrent engineering can lead to less expensive, more manufacturable designs using fewer parts. Applying these elements should go a long way to eliminating a situation in which design problems are recreated with startling regularity.

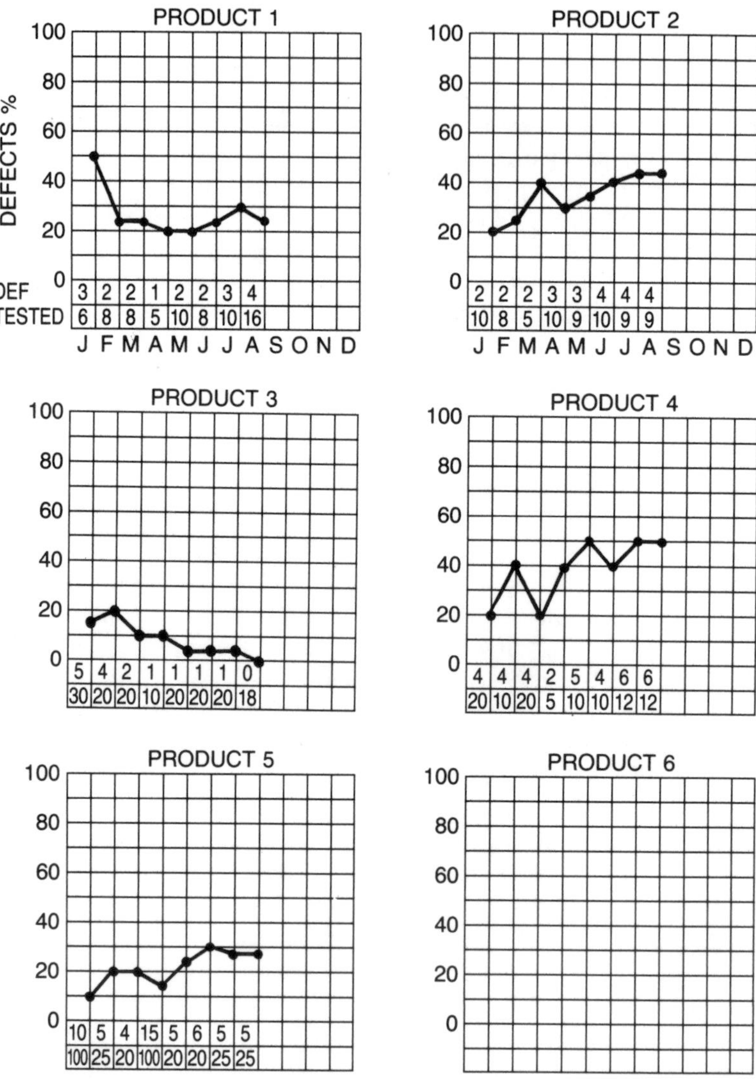

Figure 6.10 Product yield information shown in graph format.

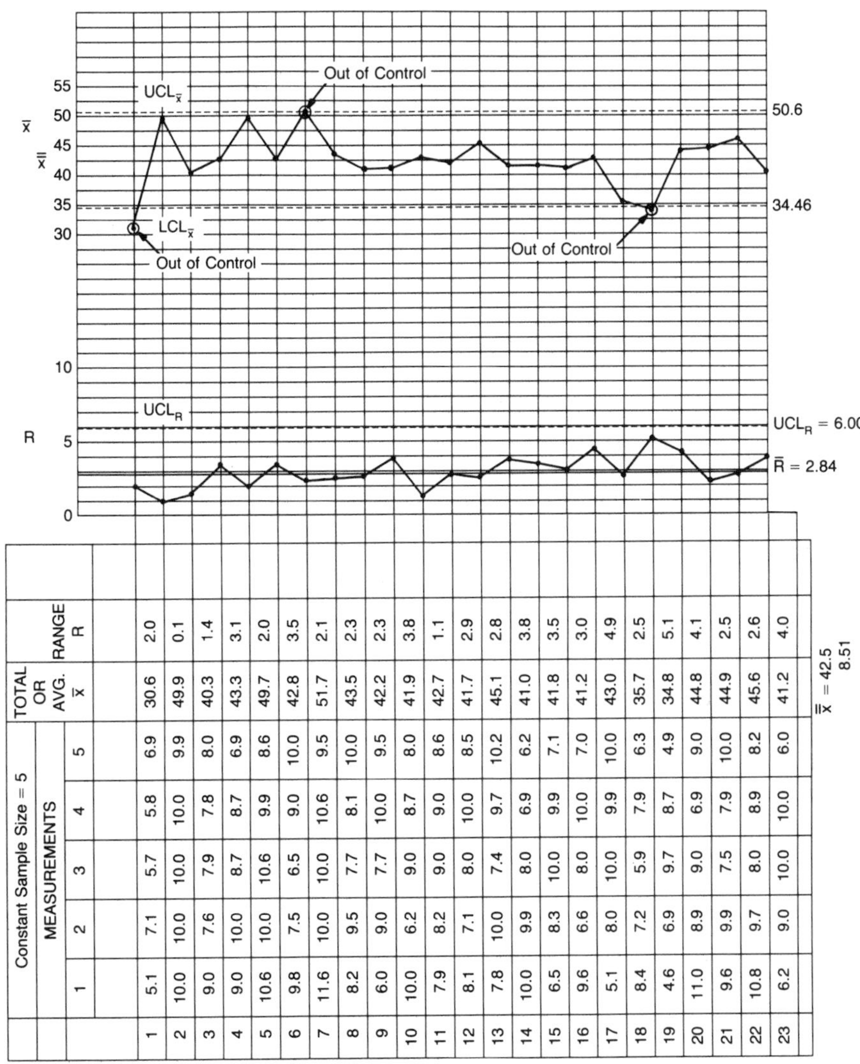

Constant Sample Size = 5

Sample	Measurements 1	2	3	4	5	Total or Avg. $\bar{X}$	Range R
1	5.1	7.1	5.7	5.8	6.9	30.6	2.0
2	10.0	10.0	10.0	10.0	9.9	49.9	0.1
3	9.0	7.6	7.9	7.8	8.0	40.3	1.4
4	9.0	10.0	8.7	8.7	6.9	43.3	3.1
5	10.6	10.0	10.6	9.9	8.6	49.7	2.0
6	9.8	7.5	6.5	9.0	10.0	42.8	3.5
7	11.6	10.0	10.0	10.6	9.5	51.7	2.1
8	8.2	9.5	7.7	8.1	10.0	43.5	2.3
9	6.0	9.0	7.7	10.0	9.5	42.2	2.3
10	10.0	6.2	9.0	8.7	8.0	41.9	3.8
11	7.9	8.2	9.0	9.0	8.6	42.7	1.1
12	8.1	7.1	8.0	10.0	8.5	41.7	2.9
13	7.8	10.0	7.4	9.7	10.2	45.1	2.8
14	10.0	9.9	8.0	6.9	6.2	41.0	3.8
15	6.5	8.3	10.0	9.9	7.1	41.8	3.5
16	9.6	6.6	8.0	10.0	7.0	41.2	3.0
17	5.1	8.0	10.0	9.9	10.0	43.0	4.9
18	8.4	7.2	5.9	7.9	6.3	35.7	2.5
19	4.6	6.9	9.7	8.7	4.9	34.8	5.1
20	11.0	8.9	9.0	6.9	9.0	44.8	4.1
21	9.6	9.9	7.5	7.9	10.0	44.9	2.5
22	10.8	9.7	8.0	8.9	8.2	45.6	2.6
23	6.2	9.0	10.0	10.0	6.0	41.2	4.0
						$\bar{\bar{x}}$ = 42.5	8.51

Figure 6.11 Typical control chart for variables (X, R).

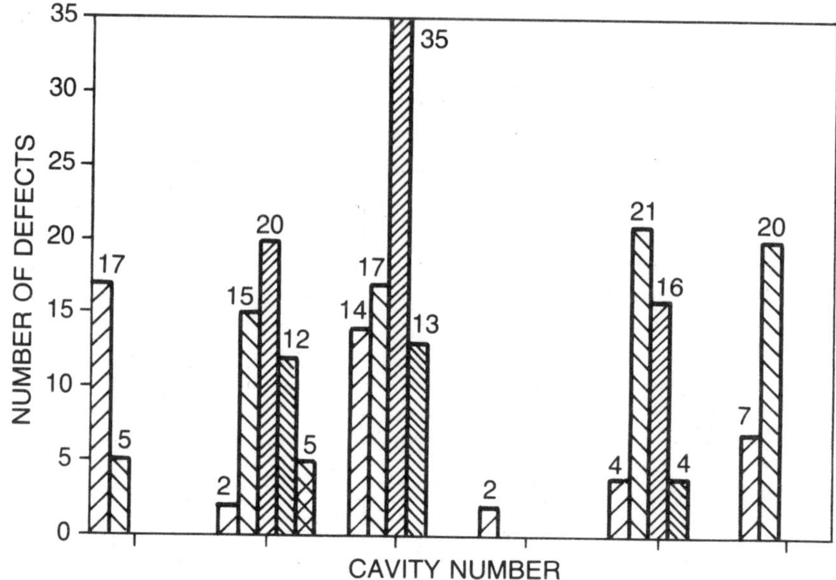

Figure 6.12 Defective (short-shot) plug analysis.

POSITION OF CAVITIES IN MOLD

1	2	3	4	5	6	7	8
9	10	11	12	13	14	15	16
17	18	19	20	21	22	23	24
25	26	27	28	29	30	31	32

DEFECTS BY CAVITY LAYOUT

	17	6	0	0	0	1	13	19
	12	6	0	0	0	0	12	17
	34	5	0	1	0	0	5	21
	17	5	0	0	0	0	8	20
TOTAL	80	22	0	0	0	1	38	77

Figure 6.13 Mold cavity layout and the distribution of defects by cavity.

Summary

The communication elements discussed are by no means complete. They are presented to provide insight into simple ways of determining what is happening in the process. Hopefully, it will benefit you by stimulating thought on how processes may vary and alert you to methods of reporting and control. Similar techniques can be used for communication to all levels of management. Establishing appropriate feedback systems and obtaining action on the results should provide true profit growth and quality improvement.

References

Besterfield, Dale H. (1986). *Quality Control,* 2nd ed., Prentice-Hall, Englewood Cliffs, N.J.

W. Frey, E. R. Ott, and E. S. Shecter, (1968). Basic concepts (of quality control), *Rutgers Conference Transactions.*

Grant, E. L., and R. S. Leavenworth (1980). *Statistical Quality Control,* 5th ed., McGraw-Hill, New York.

Juran, J. M. (1988). *Quality Control Handbook,* 4th ed., McGraw-Hill, New York.

7 • Control Chart Fundamentals

Introduction

One of the principal tools of a statistical process control program is the control chart. In this chapter we look at explain how control charts are used and what can be done with them to improve process yields and reduce nonconformance. Control charts are used in conjunction with processes to communicate to the user information about the process so that appropriate action can be taken. This chapter should make you sufficiently familiar with the use of control charts so that you can determine whether their use can be beneficial. There are two basic types of control chart, the control chart for variables in which measurements are taken and the control chart for attributes, for a go–no go decision. The variables chart plots both the average, $\overline{X}$, and the range, R, or the standard deviation, σ (sigma), directly below each value (see Fig. 7.1). The attributes chart plots nonconformances in the form of percent nonconformance (see Fig. 7.2), number of defectives, number of defects, or number of defects per unit. Other types of control charts are also available, such as cumulative sum (CuSum) charts and the exponentially weighted moving average (EWMA) chart, which are useful to describe

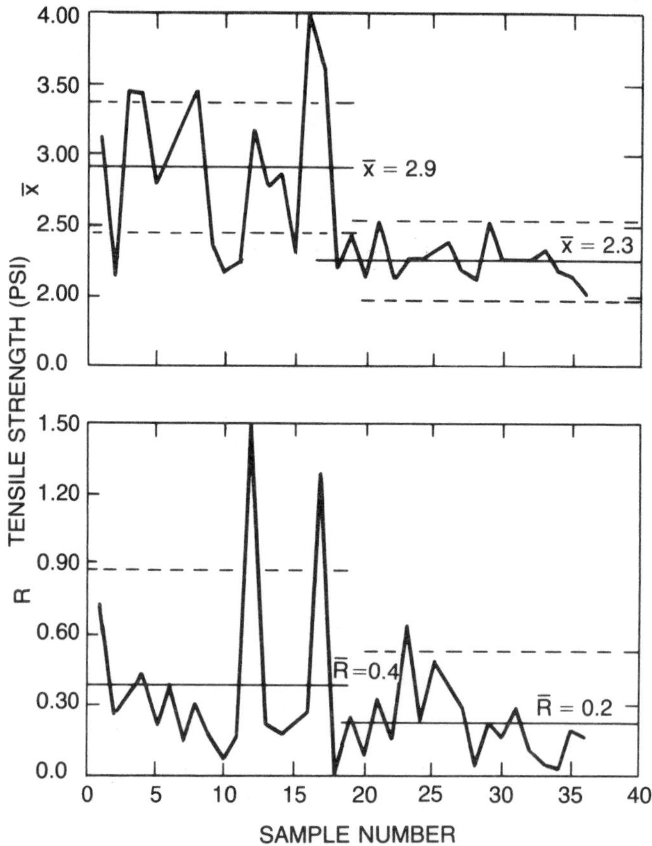

Figure 7.1 Variables control chart (uses $\overline{X}$, and R).

processes whose current levels may be influenced by earlier performance, as in chemical processes.

The Nature of Processes

Since we are going to be discussing processes, a word as to the definition of a process is in order. A simplified description of a *process* is that it is a series of events resulting in a product or service. A process converts one form of information or product into another form. It has inputs and outputs. It adds a series of events that result in this conversion. One of the inherent facts about processes is that they vary, and knowing what this variability is, and its amount, enables control. The behavior of a process is subject to a wide variety of random and non-

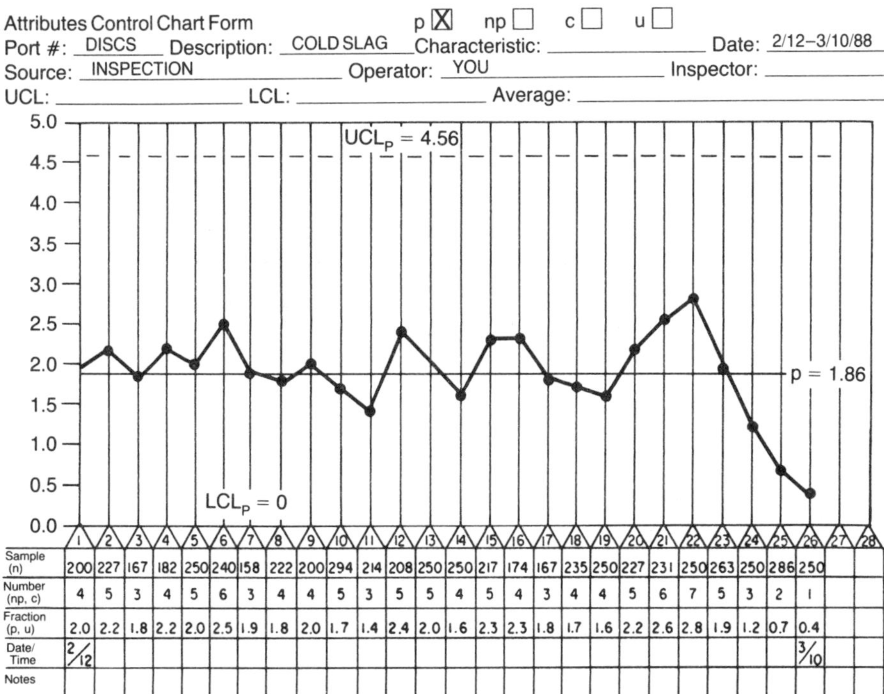

	1	2	3	4	5	6	7	8	9	10	11	12	13	14	15	16	17	18	19	20	21	22	23	24	25	26	27	28
Sample (n)	200	227	167	182	250	240	158	222	200	294	214	208	250	250	217	174	167	235	250	227	231	250	263	250	286	250		
Number (np, c)	4	5	3	4	5	6	3	4	4	5	3	5	5	4	5	4	3	4	4	5	6	7	5	3	2	1		
Fraction (p, u)	2.0	2.2	1.8	2.2	2.0	2.5	1.9	1.8	2.0	1.7	1.4	2.4	2.0	1.6	2.3	2.3	1.8	1.7	1.6	2.2	2.6	2.8	1.9	1.2	0.7	0.4		
Date/Time	2/12																									3/10		
Notes																												

Figure 7.2 Control chart for percent nonconforming.

random causes that create this variability. These random factors can include machine variations (from bearings, vibration, eccentricity), tool variations (cutting edge, shape, hardness, wear), material variations (hardness, inclusions, temper), operator variations, environmental variations, and a host of others. Taken as a whole, all the random factors contribute to the natural variability of the process. This natural variability can be determined through measurements and is measured in units of standard deviation called sigma, σ. The process capability is considered to be $\pm 3\sigma$ around the average, $\overline{X}$.

Nonrandom factors include such items as changes in materials, different machines, different heads on the same machine, ambient temperature, time variations, pressure, process temperature, chemical concentrations, operators, and others (including unique behavior of the random factors). These are some of the factors in processes that are reflected in out-of-control points on control charts. A process is out of control when it is influenced by these nonrandom factors. It defies prediction and behaves as shown in Fig. 7.3. A process is in a state of

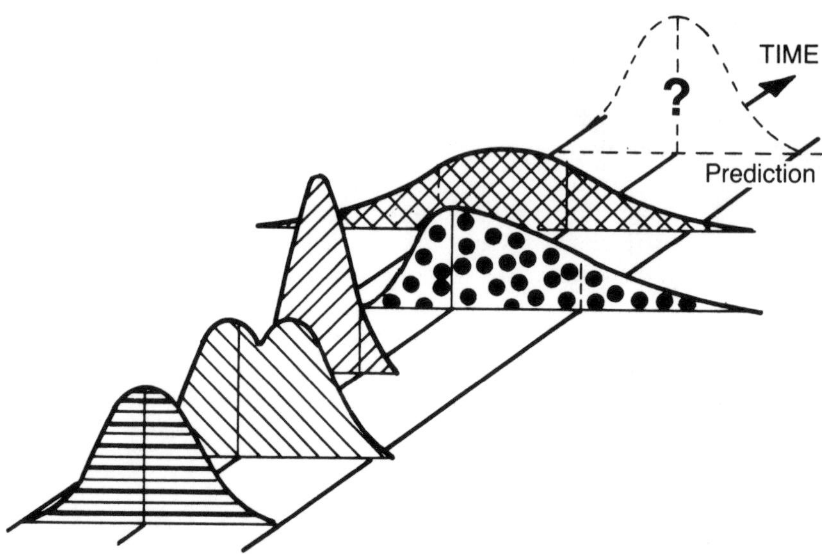

Figure 7.3 Process behavior subject to nonrandom cause. If assignable causes are present, the process is not stable and its performance is unpredictable.

Figure 7.4 Process behavior subject to nonrandom causes. If there is only random variation in the process, the output forms a stable, predictable process.

statistical control when it is influenced only by random factors. This type of process is predictable and is illustrated in Fig. 7.4. The control chart can differentiate between random and nonrandom events, so its proper usage is vital in identifying when a nonrandom event occurs. There is no other simple way to determine this deviation in the process.

For a service situation, the random and nonrandom factors are primarily people. But other factors also exist: lighting, heat, drafts, noise, and distractions; systems and procedures definitions and their interpretation vary from one system or procedure to another; physical factors, such as word processors, automatic teller machines, telefaxes, and other office equipment; management attitudes; and communications. The distinction between randomness and nonrandomness is made by the control chart, and process analysis may determine the specific cause.

Theory Underlying Control Charts

The *normal* or bell-shaped curve has very specific probabilities that are related to the average, $\overline{X}$, and the standard deviation, σ (see Fig. 7.5). When these values are known, we can predict probabilities associated with the process and the percent of conforming products that it can produce. This is true regardless of the values of $\overline{X}$ and σ when the pro-

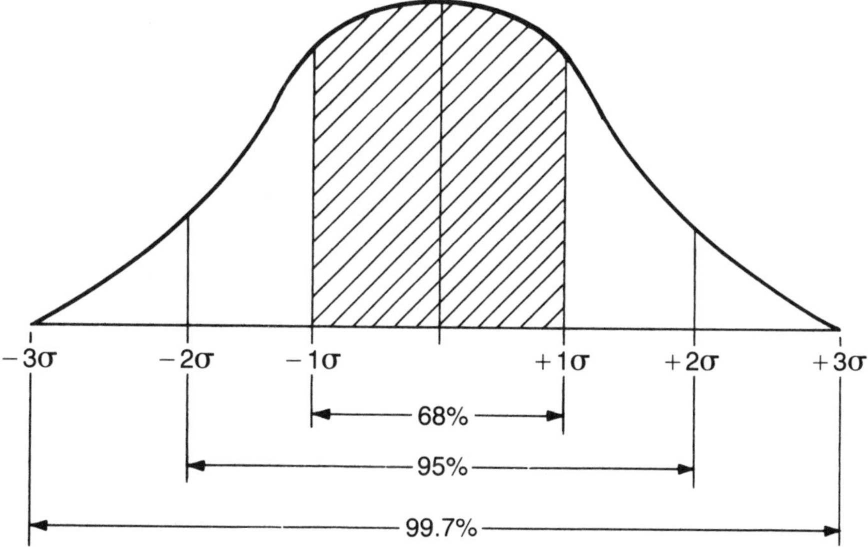

Figure 7.5 Areas under the normal curve.

cess is in statistical control. But suppose that the spread is wide rather than narrow. When measuring the points along the $\overline{X}$ axis in units of σ, the areas under the curve will be the same regardless, as shown in Fig. 7.6. If you feel uncomfortable about this concept, consider the analogy of a circle. Whether the circle is large or small, the area in a 90-degree central angle is still 25% of the total area of the circle. Another way of saying this is that there is a 25% probability of a value lying in that 90-degree segment (see Fig. 7.7). Similarly, the probability of a value (or reading) falling within $\pm 1\sigma$ is about 68.26%; $\pm 2\sigma$, about 95.45%; and $\pm 3\sigma$, 99.73%. So with the normal (bell-shaped) distribution, when the average and the standard deviation are known, the probabilities associated with values (or products) lying within specifications can be predicted. This is true whether the σ value is large or small. The areas under the curve are tabulated in most books on quality control. See Table 7.1 for areas under the normal curve.

But what about processes whose individual values are not bell-shaped? It is an interesting phenomenon that when averages are used, the distribution of these averages is close enough to normality so that the probability rules governing the normal distribution apply. Thus the control chart for averages works properly. The control chart limits are still based on 3σ limits for averages. Figure 7.8 shows how control limits are represented when distributions for individuals are shown together with distributions for averages. These limits are set at $\pm 3\sigma$ of averages and are designated $3\sigma_{\overline{X}}$. The control limits for ranges, which measure variability, are set at limits around the average range, $\overline{R}$.

We have just raised an interesting issue. The distribution of individual values is $\overline{X} \pm 3\sigma$ and the distribution for averages is $\overline{X} \pm 3\sigma_{\overline{X}}$.The exact relationship between $\sigma_{\overline{X}}$ and σ is $\sigma_{\overline{X}} = \sigma/\sqrt{n}$ where n is the sample size. Thus for a subgroup sample of 4, the spread of the averages is exactly one-half of the spread for individual items. See Fig. 7.8 for this representation; notice that control limits are set at $\pm 3\sigma_{\overline{X}}$ (3σ of the averages). By knowing the control limits, we can predict the process limits! In fact, by knowing the average range, $\overline{R}$, we can predict the process limits by dividing the $\overline{R}$ by a factor d_2, shown in Table 7.2.

Control limits of 3σ are practical because we are seeking a timely signal of a process shift. It is not good enough for individual values that are related to specification limits (not control limits) in a world-class organization because $\pm 3\sigma$ contains 99.73% of all values, thereby allowing 2.7 nonconformances per thousand or 2700 nonconformances per million [or parts per million, (ppm)]. World-class quality demands

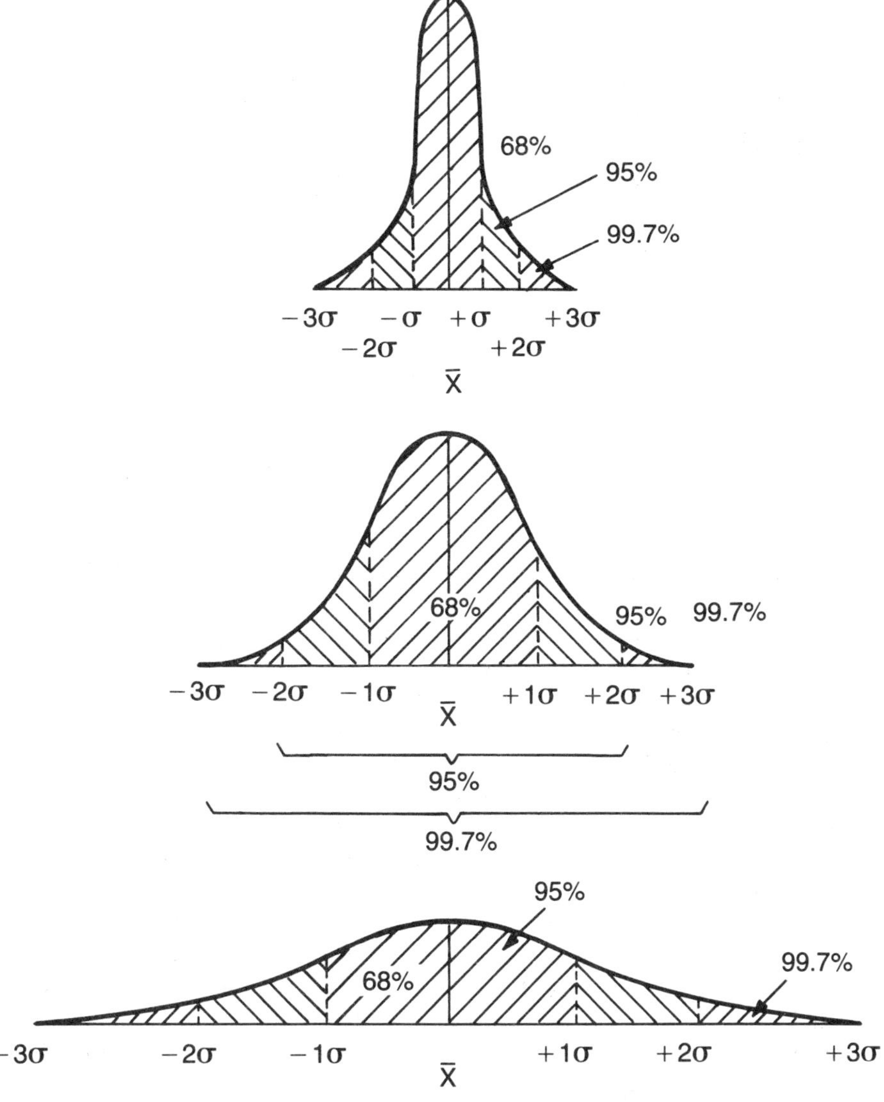

Figure 7.6 Narrow and wide distributions. The normal distribution has the same probabilities regardless of the value of sigma.

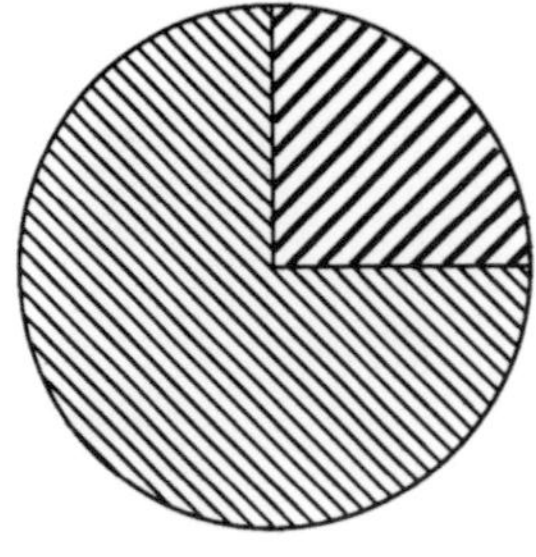

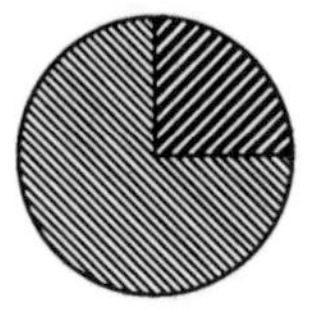

The darker shaded area is 25% of the total area in this circle.

The darker shaded area is also 25% of the shaded area even though the circle is smaller in diameter.

Figure 7.7 The area of a segment expressed as a percentage of the total area of a circle is the same for a 90° central angle, regardless of the diameter of the circle.

defect rates of less than 50 nonconformances per million or better and these rates are continually lowering.

Note that in our discussion of control limits compared to process limits, we did not discuss specification limits. The reason for this is that control limits are based on averages and the process limits are based on individual values. The process may not be related to the specification. We would like the process to be centered on the nominal value of the specification, but this does not have to be the case. To determine the potential yield of the process, we must consider the relationship of the process average and spread, $\overline{X}$ and σ, with the specification limits to determine the performance. The process or population of the product can be less than, wider than, or equal to the specification limits. The population can be centered or off-center. The control chart enables you to determine the standard deviation for individual pieces using the formula $\overline{R}/d_2$, where d_2 is dependent on the subgroup size. So a prediction of process performance (number of nonconformances produced) can also be made using the control chart provided the underlying distribution of individual values is normally distributed (see Figure 7.9).

Process Capability

The amount of this variability may be acceptable if the specification limits are wider than the natural variability of the process; it may be

Table 7.1 Areas Under the Normal Curve

Proportion of total area under the curve to the left of a vertical line drawn at $\overline{X} + Z\sigma$, where Z represents any desired value from $Z = 0$ *to* $Z = \pm 3.9$:

Z	0.09	0.08	0.07	0.06	0.05	0.04	0.03	0.02	0.01	0.00
−3.0	0.00100	0.00104	0.00107	0.00111	0.00114	0.00118	0.00122	0.00126	0.00131	0.00135
−2.9	0.0014	0.0014	0.0015	0.0015	0.0016	0.0016	0.0017	0.0017	0.0018	0.0019
−2.8	0.0019	0.0020	0.0021	0.0021	0.0022	0.0023	0.0023	0.0024	0.0025	0.0026
−2.7	0.0026	0.0027	0.0028	0.0029	0.0030	0.0031	0.0032	0.0033	0.0034	0.0035
−2.6	0.0036	0.0037	0.0038	0.0039	0.0040	0.0041	0.0043	0.0044	0.0045	0.0047
−2.5	0.0048	0.0049	0.0051	0.0052	0.0054	0.0055	0.0057	0.0059	0.0060	0.0062
−2.4	0.0064	0.0066	0.0068	0.0069	0.0071	0.0073	0.0075	0.0078	0.0080	0.0082
−2.3	0.0084	0.0087	0.0089	0.0091	0.0094	0.0096	0.0099	0.0102	0.0104	0.0107
−2.2	0.0110	0.0113	0.0116	0.0119	0.0122	0.0125	0.0129	0.0132	0.0136	0.0139
−2.1	0.0143	0.0146	0.0150	0.0154	0.0158	0.0162	0.0166	0.0170	0.0174	0.0179
−2.0	0.0183	0.0188	0.0192	0.0197	0.0202	0.0207	0.0212	0.0217	0.0222	0.0228
−1.9	0.0233	0.0239	0.0244	0.0250	0.0256	0.0262	0.0268	0.0274	0.0281	0.0287
−1.8	0.0294	0.0301	0.0307	0.0314	0.0322	0.0329	0.0336	0.0344	0.0351	0.0359
−1.7	0.0367	0.0375	0.0384	0.0392	0.0401	0.0409	0.0481	0.0427	0.0436	0.0446
−1.6	0.0455	0.0465	0.0475	0.0485	0.0495	0.0505	0.0516	0.0526	0.0537	0.0548
−1.5	0.0559	0.0571	0.0582	0.0594	0.0606	0.0618	0.0630	0.0643	0.0655	0.0668
−1.4	0.0681	0.0694	0.0708	0.0721	0.0735	0.0749	0.0764	0.0778	0.0793	0.0808
−1.3	0.0823	0.0838	0.0853	0.0869	0.0885	0.0901	0.0918	0.0934	0.0951	0.0968
−1.2	0.0985	0.1003	0.1020	0.1038	0.1057	0.1075	0.1093	0.1112	0.1131	0.1151
−1.1	0.1170	0.1190	0.1210	0.1230	0.1251	0.1271	0.1292	0.1314	0.1335	0.1357
−1.0	0.1379	0.1401	0.1423	0.1446	0.1469	0.1492	0.1515	0.1539	0.1562	0.1587
−0.9	0.1611	0.1635	0.1660	0.1685	0.1711	0.1736	0.1762	0.1788	0.1814	0.1841
−0.8	0.1867	0.1894	0.1922	0.1949	0.1977	0.2005	0.2033	0.2061	0.2090	0.2119
−0.7	0.2148	0.2177	0.2207	0.2236	0.2266	0.2297	0.2327	0.2358	0.2389	0.2420
−0.6	0.2451	0.2483	0.2514	0.2546	0.2578	0.2611	0.2643	0.2676	0.2709	0.2743
−0.5	0.2776	0.2810	0.2843	0.2877	0.2912	0.2946	0.2981	0.3015	0.3050	0.3085
−0.4	0.3121	0.3156	0.3192	0.3228	0.3264	0.3300	0.3336	0.3372	0.3409	0.3446
−0.3	0.3483	0.3520	0.3557	0.3594	0.3632	0.3669	0.3707	0.3745	0.3783	0.3821
−0.2	0.3859	0.3897	0.3936	0.3974	0.4013	0.4052	0.4090	0.4129	0.4168	0.4207
−0.1	0.4247	0.4286	0.4325	0.4364	0.4404	0.4443	0.4483	0.4522	0.4562	0.4602
−0.0	0.4641	0.4681	0.4721	0.4761	0.4801	0.4840	0.4880	0.4920	0.4960	0.5000

(continued)

Table 7.1 *(continued)*

Z	0.00	0.01	0.02	0.03	0.04	0.05	0.06	0.07	0.08
+0.0	0.5000	0.5040	0.5080	0.5120	0.5160	0.5199	0.5239	0 .5279	0.5319
+0.1	0.5398	0.5438	0.5478	0.5517	0.5557	0.5596	0.5636	0 .5675	0.5714
+0.2	0.5793	0.5832	0.5871	0.5910	0.5948	0.5987	0.6026	0 .6064	0.6103
+0.3	0.6179	0.6217	0.6255	0.6293	0.6331	0.6368	0.6406	0 .6443	0.6480
+0.4	0.6554	0.6591	0.6628	0.6664	0.6700	0.6736	0.6772	0 .6808	0.6844
+0.5	0.6915	0.6950	0.6985	0.7019	0.7054	0.7088	0.7123	0 .7157	0.7190
+0.6	0.7257	0.7291	0.7324	0.7357	0.7389	0.7422	0.7454	0 .7486	0.7517
+0.7	0.7580	0.7611	0.7642	0.7673	0.7704	0.7734	0.7764	0 .7794	0.7823
+0.8	0.7881	0.7910	0.7939	0.7967	0.7995	0.8023	0.8051	0 .8079	0.8106
+0.9	0.8159	0.8186	0.8212	0.8238	0.8264	0.8289	0.8315	0 .8340	0.8365
+1.0	0.8413	0.8438	0.8461	0.8485	0.8508	0.8531	0.8554	0 .8577	0.8599
+1.1	0.8643	0.8665	0.8686	0.8708	0.8729	0.8749	0.8770	0 .8790	0.8810
+1.2	0.8849	0.8869	0.8888	0.8907	0.8925	0.8944	0.8962	0 .8980	0.8997
+1.3	0.9032	0.9049	0.9066	0.9082	0.9099	0.9115	0.9131	0 .9147	0.9162
+1.4	0.9192	0.9207	0.9222	0.9236	0.9251	0.9265	0.9279	0 .9292	0.9306
+1.5	0.9332	0.9345	0.9357	0.9370	0.9382	0.9394	0.9406	0 .9418	0.9429
+1.6	0.9452	0.9463	0.9474	0.9484	0.9495	0.9505	0.9515	0 .9525	0.9535
+1.7	0.9554	0.9564	0.9573	0.9582	0.9591	0.9599	0.9608	0 .9616	0.9625
+1.8	0.9641	0.9649	0.9656	0.9664	0.9671	0.9678	0.9686	0 .9693	0.9699
+1.9	0.9713	0.9719	0.9726	0.9732	0.9738	0.9744	0.9750	0 .9756	0.9761
+2.0	0.9773	0.9778	0.9783	0.9788	0.9793	0.9798	0.9803	0 .9808	0.9812
+2.1	0.9821	0.9826	0.9830	0.9834	0.9838	0.9842	0.9846	0 .9850	0.9854
+2.2	0.9861	0.9864	0.9868	0.9871	0.9875	0.9878	0.9881	0 .9884	0.9887
+2.3	0.9893	0.9896	0.9898	0.9901	0.9904	0.9906	0.9909	0 .9911	0.9913
+2.4	0.9918	0.9920	0.9922	0.9925	0.9927	0.9929	0.9931	0 .9932	0.9934
+2.5	0.9938	0.9940	0.9941	0.9943	0.9945	0.9946	0.9948	0 .9949	0.9951
+2.6	0.9953	0.9955	0.9956	0.9957	0.9959	0.9960	0.9961	0 .9962	0.9963
+2.7	0.9965	0.9966	0.9967	0.9968	0.9969	0.9970	0.9971	0 .9972	0.9973
+2.8	0.9974	0.9975	0.9976	0.9977	0.9977	0.9978	0.9979	0 .9979	0.9980
+2.9	0.9981	0.9982	0.9983	0.9983	0.9984	0.9984	0.9985	0 .9985	0.9980
+3.0	0.99865	0.99869	0.99874	0.99878	0.99882	0.99886	0.0 .99889	0.99893	0.9989

Source: E. L. Grant and R. L. Leavenworth, *Statistical Quality Control*, 4th ed., McGraw-Hill, New York, 1972, pp. 642–43.

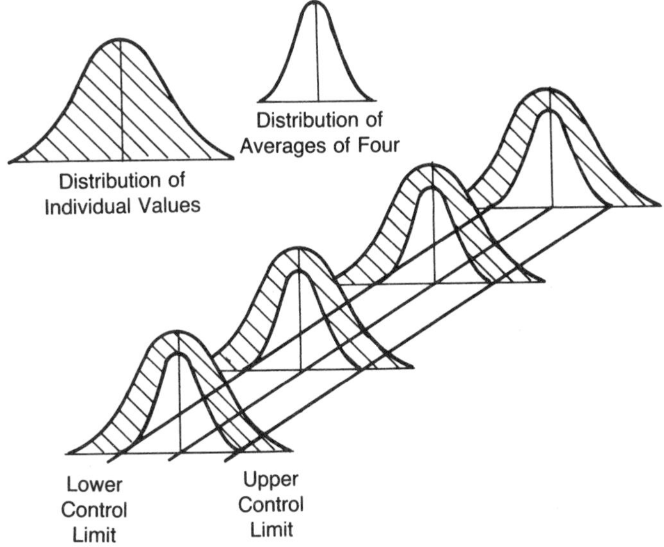

Figure 7.8 Distributions of individual values and corresponding distributions of averages. Notice there are no specifications anywhere around.

unacceptable if the natural variability exceeds the specification limits. A process whose spread is less than the specification limits is said to be *capable*. The total spread of a process is usually considered to be $\pm 3\sigma$, or a total of 6σ. This is called the *process capability*. If, in fact, the process capability is equal to the specification, 2700 defects per million items will be nonconforming. The ratio between the process capability and the specification limits is called C_p (see Fig. 7.10). An acceptable but marginal C_p is 1. A value larger than 1 indicates that the process is

Table 7.2 Factors for Computing Control Limits

n	A_2	d_2	D_3	D_3	A_3	B_4	B_3	c_4
2	1.880	1.128	0	3.267	2.659	3.267	0	0.7979
3	1.023	1.693	0	2.574	1.954	2.568	0	0.8862
4	0.729	2.059	0	2.282	1.628	2.266	0	0.9213
5	0.577	2.326	0	2.114	1.427	2.089	0	0.94
6	0.483	2.534	0	2.004	1.287	1.97	0.03	0.9515
7	0.419	2.704	0.076	1.924	1.182	1.882	0.118	0.9594

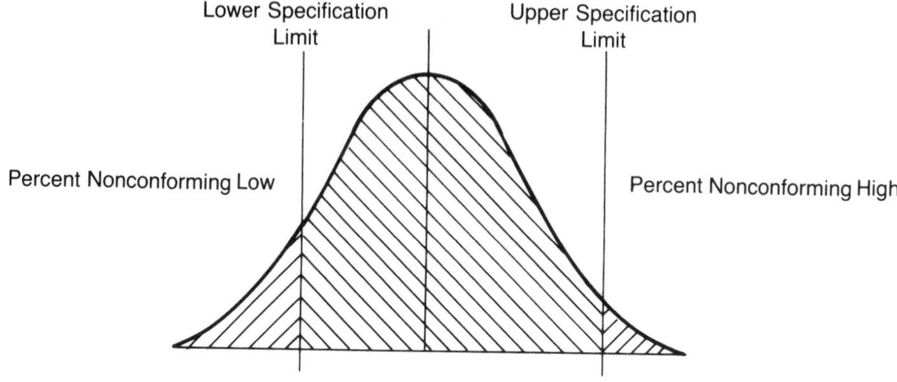

Figure 7.9 Distribution of individual values.

capable of producing larger percentages of product within tolerance. Table 7.3 shows ppm nonconformance levels for some values of C_p.

It is clear why some companies insist on values of C_p of 1.33 or values of 6σ. In Motorola's view, 6σ initially allows the distribution to shift by 1.5σ, resulting in a 4.5σ performance, or 3.4 ppm. Ultimately, this shift in average will be eliminated, resulting in a process capable of producing nonconforming levels of 1 part per billion. A C_p below 1 indicates that there will be a higher level of nonconformances, necessitating some form of testing or inspection to provide full compliance. This adds additional and unnecessary cost, resulting in lower profits, longer delivery times, and generally poorer quality (see Chapter 5 for more details).

To make sure that the process average is not too close to one specification limit, another term, designated C_{pk}, is used. This measures the ratio of the distance between 30 and the absolute value of the nearest specification limit and the average (see Fig. 7.11). This value should also exceed 1. Table 7.3 applies here as well.

Using a Control Chart for Process Control

If control charts are not used, an indication of out-of-control conditions may be the generation of nonconforming products. This is intolerable for world-class competition. When this occurs it is essential to introduce corrective measures, and one of the best ways to determine what to do is to introduce a control chart. This will help determine the magnitude of the problem and provide insight as to how to proceed. It is also desirable to prepare a frequency distribution or histogram (see

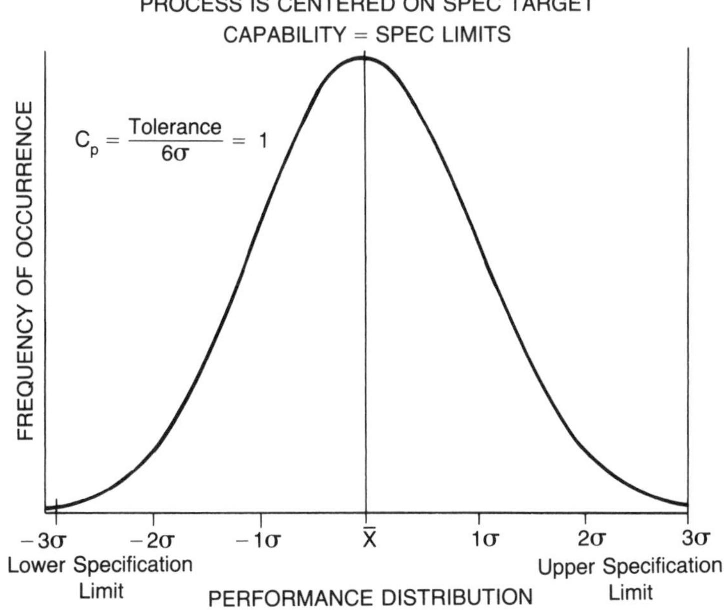

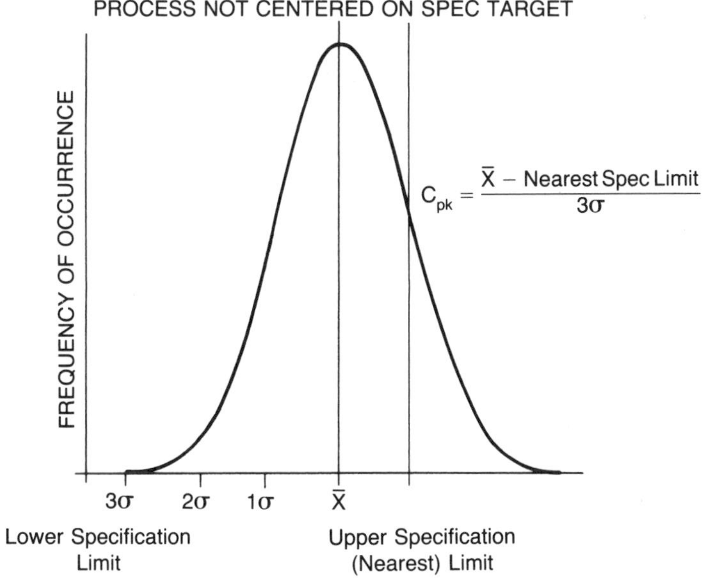

Figure 7.10 C_p and C_{pk} illustrations.

Table 7.3 Comparison of σ with c_p, ppm, and ppb

σ	c_p	*ppm*	*ppb*
3.0	1.00	1350	
3.5	1.166	232	
4.0	1.33	31.7	
4.5	1.66	3.4	
5.0	1.83		287
5.5	1.83		19
6.0	2.00		1

Fig. 7.11) to get a visual representation of process behavior when variables data can be obtained. (Sometimes only attribute data can be obtained; see Chapters 6 and 10 for some ideas on troubleshooting in such cases.) The control chart can then be used by engineers or others knowledgeable in the process to determine how the process is behaving and thus provide innovative solutions.

Out-of-control points signal an assignable cause whose source should be sought with great diligence. Finding these causes and eliminating them are the key to bringing a process under control. Points that are out of control on the average chart indicate shifts in level. Points that are out of control on the range (or σ) chart indicate shifts in spread. Both provide clues to the causes. The timeliness of the discovery is a key to identifying and eliminating assignable causes and bringing the process under control.

When a process is not capable, and under a state of control, the spread or standard deviation (σ) is too large for the tolerance and ways must be found to shrink the spread or widen the tolerance. Otherwise, nonconformances will be produced and 100% inspection or test will have to be performed. The increased cost and lower quality levels in the product stream cannot be tolerated. To search for ways to reduce the spread, changes must be introduced. It may be necessary to tighten processing parameters, run statistically designed experiments (see Chapter 10), or take other corrective actions. For example, a machine may need bearings, may require isolation from vibration transmitted by adjacent machinery, or a new machine may be required. Statistical methods can help determine what is necessary to reduce variability and often these methods, discussed in Chapters 8 and 10, can reduce the need for capital investment. Capital investments should consider the cost of nonconformance as well as the increased productivity that new equipment usually provides.

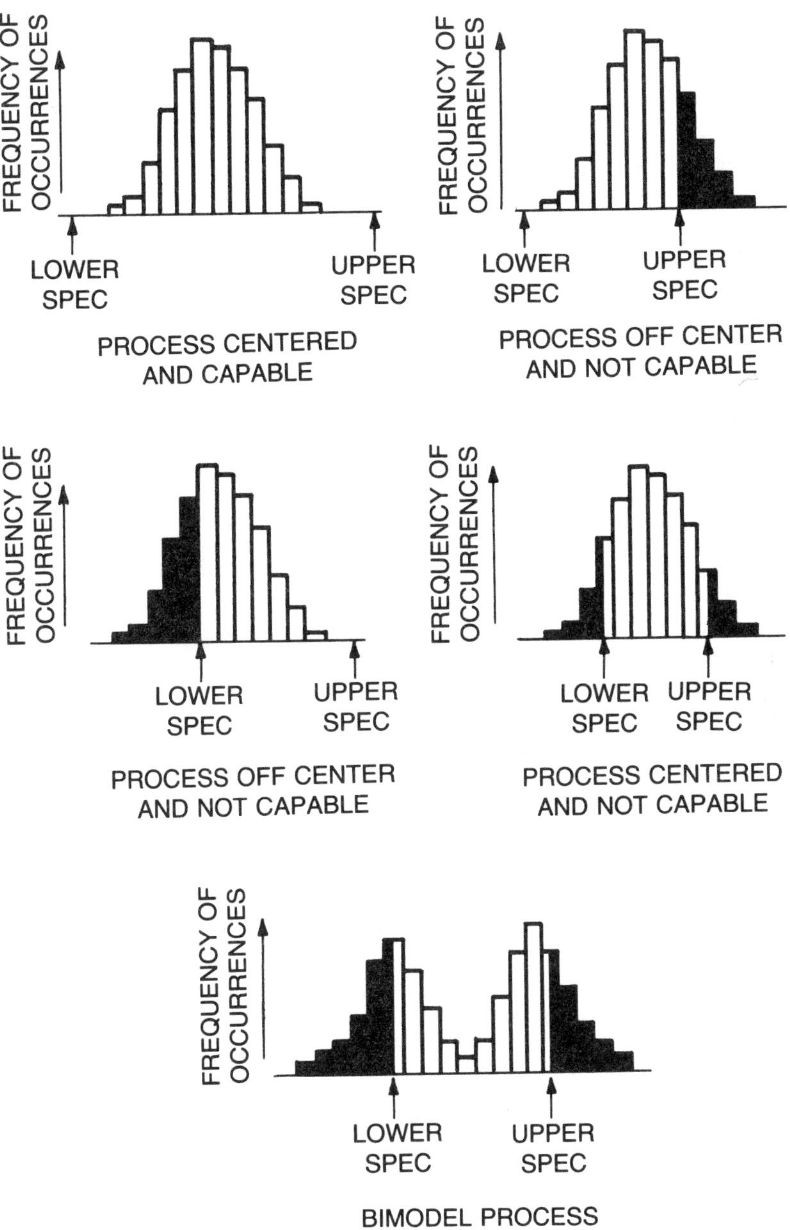

Figure 7.11 Histograms.

Processes behave in many ways. Some processes are subject to sudden shifts in level, some to gradual changes up or down, some to cyclical changes, some to sudden shifts in variability, some to gradual shifts in variability, and others to causes too numerous to mention. The astute engineer, manager, or operator must be aware of these possibilities to control them effectively and to detect quickly causes of change. The control chart is a valuable tool to provide guidance in this area (see Tables 7.4 and 7.5 for routine steps in constructing a control chart for variables). Be aware that the application is not routine and care must be exercised as to where, when, and how control charts for variables should be applied. The following case history is an actual example of the misuse of control charts by skilled engineers and statisticians and subsequent correction of their use, enabling major process improvements.

Case History: Nickel–Cadmium Battery Plaque Manufacture

An example of control chart usage illustrating some of the factors to be considered occurred in the manufacture of battery plaques for nickel–cadmium batteries. One operation involves "sizing" the plaques in a floating platen hydraulic press to provide uniform plaque thickness. A plaque made of sintered nickel on a nickel mesh is about 8 in. by 10 in. by 0.035 in. thick. Since these plaques are subsequently cut to smaller plates and the plates stacked (much like a deck of cards) with separators between each plate to form a battery cell, plaque thickness is an important control characteristic.

The quality engineer established a control chart that used a sample of five plaques and measured the thickness at the center of the plaque, from which the average and range were determined. The process ran nicely within control at ± 0.006 natural tolerance. Unfortunately, the specification required a tolerance of ± 0.001. Since production had to continue, a 100% sorting operation was set up to allow battery cells to be selectively assembled. Although this added to the cost, it was necessary to maintain production. The process defeated many attempts to reduce its natural variability and the engineering staff felt that the uncontrollable springback in the sintered plaques after the sizing operation was the major cause of the problem and had to be tolerated.

The quality engineer decided to try a different sampling method for control chart purposes. He coded the rectangular plaque so that it was always inserted in the hydraulic sizing press in the same orientation. Then he measured each of five plaques in five places: upper left, upper right, lower left, lower right, and center. He then maintained a control

Table 7.4 Steps for Variables Control Chart Contruction

1. Select a process.
2. Select a characteristic.
3. Determine how to get a sample.
4. Take a sample of 3, 4, 5, or 6 pieces.
5. Determine average, $\overline{X}$, and range, $\overline{R}$.
6. Plot $\overline{X}$ on a graph.
7. Plot $\overline{R}$ on the graph directly under $\overline{X}$.
8. Collect 10 to 20 samples as in step 4.
9. Plot each sample as it is obtained.
10. Determine the grand average, $\overline{X}$, and the average range, $\overline{R}$.
11. Calculate control limits for the average by $A_2\overline{R}$ and add and subtract the product to and from $\overline{X}$. Thus $\overline{X} \pm A_2\overline{R}$.
12. Calculate control limits for the range, $D_3\overline{R}$ lower limit, and $D_4\overline{R}$ upper limit.
13. Examine the control charts for evidence of control or lack thereof.
14. Take action on out-of-control conditions.
15. Calculate the process capability: $\frac{\overline{R}}{d_2} = \sigma$; 3σ = This is considered the process capability.
16. Compare $\overline{X} \pm 3\sigma$ with the specification limits.
17. Determine whether the process is capable of meeting specifications.
18. If the process performs well within the specs, consider whether the chart should be deleted or whether the specs should be tightened.
19. If the process performs outside the specs, get help from other engineers.
20. Work toward reducing process validity.

Notes: 1. Step 3 is very important since it will determine $\overline{R}$, on which all calculations and decisions are based.
2. Wherever $\overline{R}$ is used, σ or s may be substituted, but then control chart calculations are changed to $\overline{X} \pm A\ _1\sigma$: $LCL_\sigma = B_3\sigma$; $UCL_\sigma = B_4\sigma$.

chart for each of the five locations. The results of several days worth of data, showed that the upper left was consistently thicker than the lower right. It appeared as though the platens in the press were not parallel. Presented with the data, the engineering staff said that an upper platen on the press was floating (on a ball joint) and therefore would align itself with the fixed lower platen, and that something else also had to be wrong. When the engineers finally agreed to make a thorough check

Table 7.5 Steps for Attributes Control Chart Construction

1. A control chart for percent defective can be used when it is desired to know what the percent defective is for daily production from a process. The characteristics measured may only be determinable as good or bad or they can be variables data measured on a go–no go basis.
2. Determine what data are desired.
3. Calculate the percent defective, p, or fraction defective (as a decimal) each day.
4. Record the number inspected or tested.
5. Plot the daily percent defective or the fraction defective on a graph for p.
6. After recording 10 to 15 points, determine the average percent defective, $\bar{p}$.
7. Calculate $\bar{p} \pm 3\sqrt{[\bar{p}(1-\bar{p})]/n}$ where n is the average number of items inspected or tested. If n varies from plot to plot, it is usally OK to use the average n or ($\bar{n}$) for the calculation. These values provide the control limits
8. Plot the control limits, thus converting the graph to a control chart.
9. For any points close to the control limits, if n is larger than $\bar{n}$, the control limits will narrow; then the point is out of control and there is an assignable cause. Look for it immediately!
10. If n is smaller than $\bar{n}$, you may recalculate the control limit as in step 7. If the point is still out of control, look for an assignable cause at once!
11. The control chart for p is used as an indicator of process stability, but in these cases we wish to improve the process below the lower control limit or obtain results consistently below $\bar{p}$. This requires rapid analysis and use of the results to force continual improvement of $\bar{p}$.
12. The basic difference between the control chart for fraction defective and the control chart for variables is that for the fraction defective chart, we do not wish to maintain control. We wish to break through the lower control limit or get consistently below the average p.

of the press they found that the floating platen press was, indeed, a fixed-platen press. Despite the fact that the company had bought and paid for a more expensive floating-platen press, the vendor had delivered a fixed-platen press. Since this had occurred several years earlier and the press vendor was now out of business, restitution was not possible.

A few simple adjustments to the upper platen alignment and the addition of shims to the press corrected the condition, however, and made possible a process capability tolerance of ±0.0008, resulting in virtually 100% conforming products. This eliminated the need to sort for selective assembly, which both saved money and improved the product. The process control method was changed to measure five samples from each of two diagonal corners of five consecutive plaques, and the problem was resolved. Two control charts were maintained, one for the thickness at each corner.

The purpose of this example is to highlight that a sample be selected before blindly starting a control chart. The sample selection process determines the range of values and since the control limits are directly proportional to the range, these limits (and out-of-control indications) determine when to seek an assignable cause. A rule of thumb is to select a sample over a period of time that is likely to include the natural variability of the process and to exclude longer-term shifts. Make sure that the sample measurements are on different items (but perhaps at the same location on each of the items), not at different locations on the same item.

For high-volume processes, consecutive pieces will generally be so close in value that the range will be too small and an excessive number of points will be out of control. The causes for this may be impossible to detect and the chart will eventually fall into disuse. In this circumstance, the sample might be every fifth or tenth piece since it provides a reasonable range to use for the chart. Alternatively, taking a sample from an increment of time such as a 5- or 10-minute period, disregarding the sequence in that interval, might be suitable. When multiple heads or multiple cavities are involved, consider taking the subgroup from one head or one cavity and keep charts on several heads or cavities (see Wheeler and Chambers, 1986).

Attribute Control Chart

Thus far, we have discussed the control chart for variables where the main goal is to keep the process within control by minimizing process shifts and eliminating assignable causes. We might add that it is desirable to reduce the range or spread and thus improve the uniformity of the product. The control chart for attributes which may be a percent defective (*p* chart), number defective (*np* chart), defects per unit (*u* chart), or number of defects (*c* chart) is a control chart used to monitor control in order to achieve changes in level. This is truly a major difference from the uses for which attributes charts are designed. A variables chart is used to maintain control; an attributes chart is used to achieve breakthrough.

Whereas the control chart for variables is based on the normal distribution, the *p* and the *np* charts are based on the binomial distribution, and the control charts for *u* or *c* are based on the Poisson distribution. Regardless, the philosophy of control charting is still the same: It is a tool to differentiate assignable, and therefore detectable, differences from those that are a part of the normal process.

Table 7.6 Inspection Results

Item	Date	Number Inspected	Number Nonconforming	Percent Nonconforming
1	Oct. 1	154	21	13.6
2	Oct. 2	123	18	14.6
3	Oct. 3	133	17	12.8
4	Oct. 4	162	34	21.0
5	Oct. 5	166	30	18.1
6	Oct. 8	143	43	30.1
7	Oct. 9	160	20	12.5
8	Oct. 10	162	44	27.1
9	Oct. 11	136	24	17.6
10	Oct. 12	135	32	23.7
11	Oct. 15	144	39	27.1
12	Oct. 16	152	40	26.3
13	Oct. 17	140	20	14.3
14	Oct. 18	149	37	24.8
15	Oct. 19	162	34	21.0
16	Oct. 22	151	20	13.2
17	Oct. 23	117	22	18.8
18	Oct. 24	160	43	26.9
19	Oct. 25	144	40	27.8
20	Oct. 26	169	38	22.4

Consider the use of these charts. The object now is to use the chart not only to detect abnormalities in a timely fashion, and therefore increase the likelihood of detecting a problem, but also to reduce the level of defects being experienced. For example, if a process is running at a 10% defect level, the desire is to reduce the percent defective to a lower level and ultimately to zero. This is done using the information related to types of defectives contributing to the overall percent defective and finding ways to eliminate the defect-causing factors. See Chapter 6 for methods of data collection to identify problems and Chapter 8 for ways to correct problems.

A control chart can help determine whether a significant difference has occurred (from a statistical point). To plot a control chart for percent defective, one merely keeps a record of percent defective values for a given time frame, such as hourly, daily, or weekly (see Table 7.6). Then a plot of these data is presented (see Fig. 7.12). After about 10 points are obtained, control limits may be calculated using the appropriate formula shown in the appendix of this chapter. Table 7.5 provides a step-by-step procedure for plotting a control chart for percent defective. Similar procedures can be used for other attribute charts.

When the number of items used for a *p* or *np* chart varies, the control limits will vary. Frequently, though to reduce the need for calculation, the average number of items is used as the factor for *n*. This is especially true since these charts are maintained to help create change rather than to keep things stable. As long as the points plotted remain within $\pm 3\sigma$ limits, no assignable causes have affected the process. When points fall outside the limits, a shift has occurred whose cause can be detected or which has been caused as the result of corrective action. In general, the same rules for out-of-control conditions apply to these types of charts as for the control charts for variables, except that a range or σ chart is not involved.

Data Collection and Corrective Action

How to collect data so that the evidence will be very visible to the user is covered in Chapter 6. At this point it must be realized that proper data collection is a key ingredient to successful management for world-class quality. Methods whereby corrective action is introduced and institutionalized are largely a function of the organization. However, some basic practices are endemic to all organizations. Alternatives for achieving corrective measures are discussed in Chapter 8.

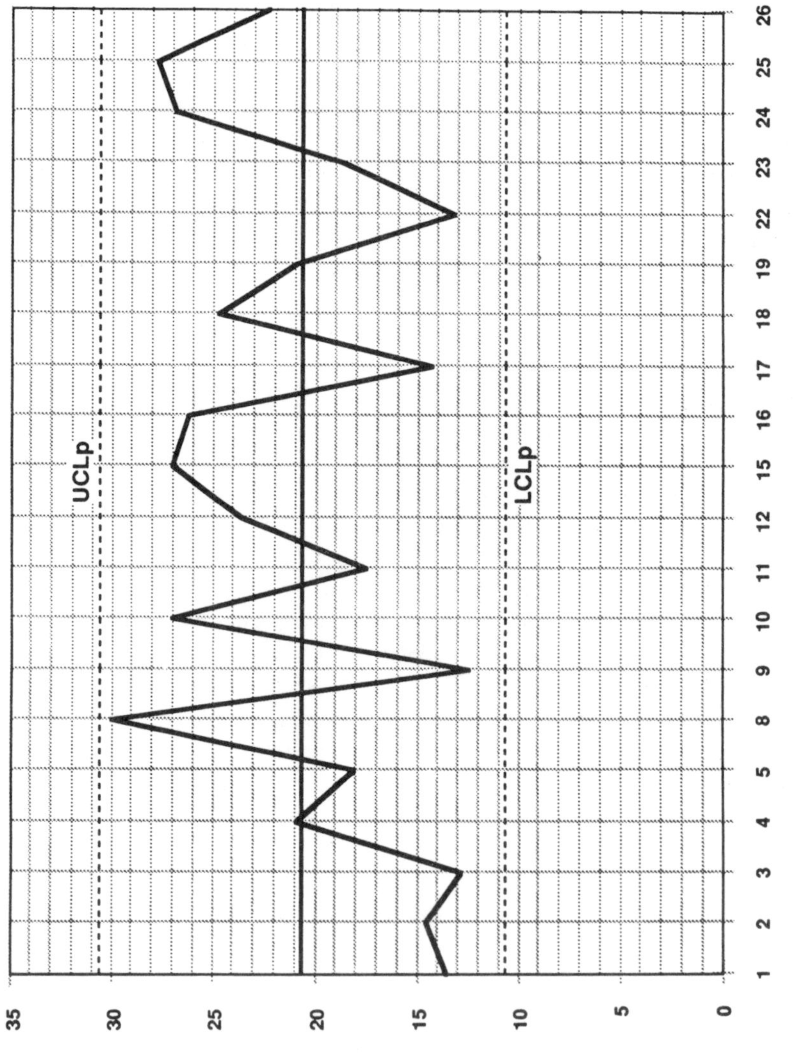
UCLp
LCLp
OCTOBER 1990
PERCENTAGE OF NONCONFORMITY

Summary

Control charts can be powerful allies against producing nonconforming products and providing guidance for efficient use of people's time. Properly used, they are major weapons in the struggle to produce defect-free products at competitive costs. Their primary use is as an indication of process control status and to provide timely warnings of process shifts. The sensitivity of the chart to shifts in performance and the timeliness of the warning are major factors in achieving better control. The control chart for variables is also very useful in determining process capability, thus enabling the determination of the process limits. By comparing the process capability to the specification limits, predictions of yields can be made and preproduction or early production changes can be designed and introduced to create higher yields. These charts also provide visibility into performance and can be used as a target for reducing variability and improving yields throughout the entire process, regardless of whether the output is a product or a service. Use of these charts is essential in managing for world-class quality since they provide a management tool for determining product yields and provide continuous pressure for improvement.

Reference

Wheeler, Donald J. and David S. Chambers, *Understanding Statistical Process Control,* (1986). Statistical Process Control A, Inc., Knoxville, Tenn.

Figure 7.12 Control chart for percent defective. The control limits are set based on an average sample size of 148. In determining out-of-control conditions, points that are near the control limits on either side should have the limit recalculated based on the actual number of items inspected. The out-of-control points are indications that some assignable cause has created this condition. By knowing when the occurence happened, the cause can be more readily determined. In actual use, the objective is to reduce the percent nonconforming through continuous improvement.

8 · Corrective Action

Introduction

To achieve world-class quality, a good quality system involving a timely and effective corrective action program is essential. Corrective action is not as simple as might appear on the surface because, to a large extent, it deals with people who often have conflicting views and priorities. The purpose of this chapter is to explore in depth what is meant by *corrective action* and what have proven to be the most effective types of corrective action programs. There are four essential ingredients in the corrective action cycle—(1) information, (2) analysis, (3) action, and (4) follow-up—and each will be examined in turn.

Information

Information, in the form of data, must be obtained such that it is timely, accurate, and relevant. It must also be presented in a manner that communicates intelligence and visibility. Reports to higher levels of management must be scannable to be useful.

Where are these data gathered? They should be a part of the normal day-to-day inspection and testing activities. In fact, the principal purpose of inspection and testing should be to collect and provide data that can be used to effect corrective action. This sounds simple enough, but most people think the major purpose of inspection and testing is to make sure that only good products pass to the next level of assembly or on to the customer. This is terribly wrong, since it influences the manner in which we work. If, in fact, the major purpose of inspection and testing is to make sure that only good products get through, rejected items may simply wait in a rework area or usually be analyzed on a first-in, first-out basis. This creates delays in knowing what is currently happening and often makes the information obsolete by the time it becomes available.

In the nonmanufacturing sector, operations may not be called inspection or testing but may be given other titles, or the do-over operations may be a part of the operator's job. If a task is performed in a certain manner and the person doing that task has been trained to do the task and keep repeating it until it comes out right, that is the way that task will be done. People tend to feel that there is no reason to question the number of times things get done over if that is the way it has always been done. This reasoning is wrong! If something has to get done over, there is a reason it did not come out right the first time. Finding that cause and eliminating it from the process will result in less redo. Reducing redo is the role of corrective action. For example, if it takes 2 weeks of each month to close the financial books, as some companies have reported, knowing how much of that effort is redoing tasks because the books do not come out right (balance), the causes for the redo must be determined and the process corrected to eliminate these factors. The time to close the books will then be reduced as each cause of error and redo is eliminated.

The first rule, then, is to collect information quickly on currently manufactured products. The rejects contain this information, so they must be analyzed promptly. For many inspection operations, the data are readily available, but for more complex operations such as chemical processes or electrical testing, some form of analysis or troubleshooting may have to be done to provide useful data.

Another rule is to collect the data in a readily usable form. In the case of process control, for example, a control chart is a popular and effective way of doing this. A process in a state of statistical control, that is, all plotted points lie within control limits, is not necessarily sufficient. An estimate of the standard deviation, s', and average, m, should show the process to have the capability of behaving well within the specification limits, and indeed, averaging close to the nominal value of the specification. This provides some latitude in allowing a drift or shift in the process without producing nonconforming products. This strategy can reduce inspection, test, and rework costs and can lead to 100% compliance—a very desirable goal.

Motorola has established a 6σ process as a requirement for all their processes. This is based on the assumption of normality and is used as a universal goal for all processes to achieve this level of performance by 1992. According to Motorola's concepts, 6σ is equal to allowing 3 defects per million opportunities or 3 parts per million (ppm). In this perception, they assume that the process will shift about the nominal value by 1.5σ. Thus, at its worst level (or extreme shift), the distribution will have 4.5σ between its average and the specification limit. This is equivalent to 3 ppm nonconformance. If the process can be held steadily at the central value, only 1-part-per-billion opportunities will

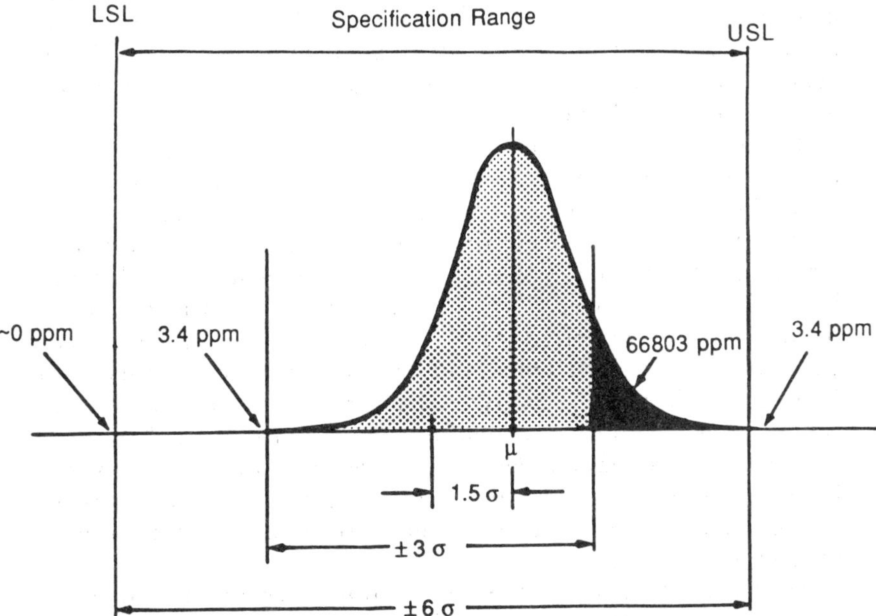

Fig 8.1 Distribution with 6σ spread showing a shift of 1.5σ, resulting in 3.4 parts per million nonconformance. If the distribution were centered on the mean, there would be about 1 nonconformance per billion.

be nonconforming. This may become a future goal. Figure 8.1 illustrates this concept.

There are two types of data that can be obtained: variables data and attributes data. The former is a measurement and is more desirable than attributes data since it provides more information. Attributes data are data indicating conformance or lack of conformance to a specification. Variables data may be collected in many forms, such as lists or tables of numbers, but this does not present a format that the eye or mind can easily integrate into a perspective of how the process is behaving.

Another form of variables data that is helpful to distinguish patterns of performance is the frequency distribution or histogram. This plot has the advantage of showing the average, spread, and shape of the characteristic distribution and is very helpful in tracking down problem causes. The problem with data of this type is that there may simply be too much information to grasp to enable intelligent corrective action. For example, suppose that an item has 50 or 100 parameters. Variables data in whatever form may simply be too much to interpret. The availability of computer analysis can help in this situation, but care must be exercised to avoid getting too much data and thwarting our ability to do anything. In these instances, and in those where variables data are not possible (e.g., acceptability of a solder joint), attributes data are extremely helpful.

The collection and display of attribute data have been poorly handled more often than not. Attribute data can best be interpreted when shown in columnar form. The examples in Fig. 8.2 and 8.3 provide a poor format and a good format, respectively. Note that in the poor format, additional analysis must be done to determine what is happening, and this takes time. When it does, the information is delayed, so the true picture is not presented in a timely fashion and the result may be too late to correct the problem before it becomes worse. In the good format, the columns of defect types enable the reader to tell at a glance what the major problems are and their relative magnitude. These data can also be displayed in a Pareto chart, which shows the defects in a bar chart format, with the largest value first, the next largest second, and so on. This type of chart helps convey the information in an easily understandable format.

The Pareto chart (Fig. 8.4) is another way to display the data to aid in understanding. The advantage of the Pareto chart is that it graphically portrays the events that it is measuring in descending order with the highest element plotted first. A properly laid out report in rows

Company X-Y-Z

Date	Product Quality Record for ABC Department			
XX	1026 units checked	1002 units good,	24 defects	16 defects class A, 8 defects class F,
XA	905 units checked	924 units good,	41 defects	1 class B, 9 class A, 20 class F, 10 class C, 1 class L
XB	1027	985	42	18 Class A defects, 1 class B, 14 class F, 15 class H
XC	1148	1126	22	12 class A, 10 class F
XD	1921	1827	94	37 defect A, 32 defect F, 4 defect G, 6 defect H, 6 defect K, 3 defect N, 1 defect O, 3 defect P
XE	986	909	77	41 defects class A, 3 defects class D, 2 defects class E, 27 defects class F, 40 defects class H
XF	1422	1400	22	6 defects class A, 16 defects class F
XG	1227	1201	26	3 defects class A, 1 class C, 20 class F, 1 class H, 1 class K

Fig 8.2 Product quality record for XYZ department: poor format for data display.

				DEFECT TYPE																			
				Major Classif.					Major Classif.					Major Classif.					Major Classif.				
Date	No Eval.	No Good	No Def.	Defect A	Defect B	Defect C	Defect D	Defect E	Defect F	Defect G	Defect H	Defect I	Defect J	Defect K	Defect L	Defect M	Defect N	Defect O	Defect P	Defect Q	Defect R	Defect S	Defect T
XX	1026	1002	24	16					8														
XA	925	924	41	9	1				20			10			1								
XB	1027	985	42	18					14		5					5							
XC	1948	1126	22	12					10														
XD	1921	1827	94	37					32	4			6	6			3	1	5				
XE	986	909	77	41			3	2	27		4												
XF	1422	1400	22	6					16														
XG	1227	1201	26	3		1			20		1		1										

and columns should not require a Pareto chart since it is only necessary to scan the rows or columns for large numbers of nonconformities to determine the nature of the problem. Nevertheless, the Pareto chart is useful for displaying all types of attribute data since it is simple to understand.

Scannability of Data

Bear in mind that the information is being collected so that status can be assessed and action can be taken. Make the reader's job easy by giving visibility because he or she does not have the time to devote to the interpretation. The objective is to get a message across quickly and succinctly.

Relevance and Accuracy of Data

Relevance and accuracy are the final two ingredients of data gathering. Although these would seem self-explanatory, a word about each is important. There are opportunities to gather so much data in the production or service industries that a critical selection must be made. If relevant data are not provided or if irrelevant data are intermingled with important data, they will all be ignored. It is better to err on the side of insufficient data than provide so much that none of it is used. Accuracy also seems to be self-explanatory, but here, too, inaccurate data not only casts doubt upon itself but on the accurate data as well. Make sure that what is presented is accurate to keep all data from being ignored.

Analysis

Having gathered the data, it is next necessary to analyze it. If the information is properly collected, analysis should be straightforward and may even be automatic by visually scanning data presented in the proper format. It is recognized this is not always possible. It may be necessary to analyze the data presented to make a decision about alter-

---

Fig 8.3 Product quality record for XYZ department: good format for data display. This is a manually generated report, with defects arranged in columns to permit easy identification of major defects.

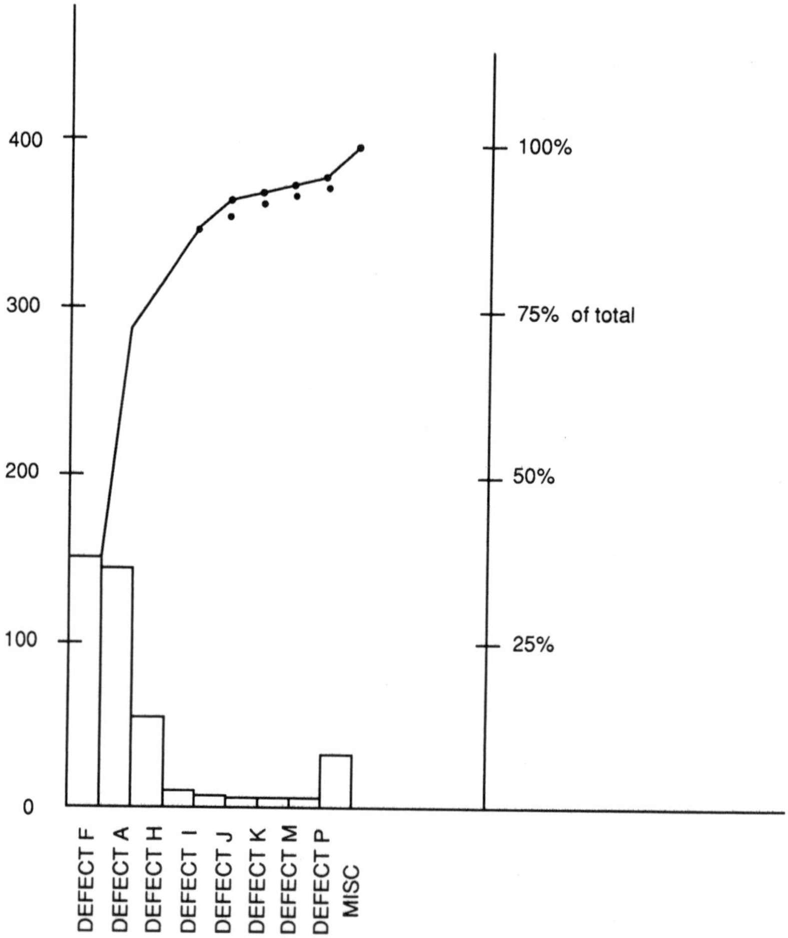

Fig. 8.4 Pareto chart.

native courses of action. On the other hand, it may be desirable to review data simply to see whether any action is warranted.

One of the most dramatic uses of simple data analysis affected the entire industry. The situation involved the manufacturing of batteries and used frequency distributions to plot battery capacity and has been explained in Chapter 6. Analysis may also require statistical tests of significance, such as analysis of variance, analysis of means, t test, χ^2 test, or some other technique. These can be done manually or by computer in order to identify an item or items in need of corrective action. Where data analysis is performed, it is almost always advisable to plot the data to see what is happening. Graphs, frequency distributions, control charts, Pareto charts, and scatter diagrams are some common ways of presenting data.

Action

Effective action is the heart of the corrective action program. There are many approaches that can be used for corrective action. First, let's define corrective action. Certainly, it means correcting the deficient product. What it must really accomplish is correcting the system or process that allowed the problem to occur in the first place—correcting the underlying cause. Frequently, there are many possible causes. Consider, for example, a defective solder joint. Some of the possible causes for this single event are:

1. Operator improperly trained
2. Part leads not properly cleaned
3. Wrong type of soldering iron used
4. Quality control inspection too critical
5. Improper or no flux used
6. Bad solder
7. Operator carelessness
8. Part movement prior to solder hardening
9. Difficult design
10. Insufficient lighting in area
11. Poor technique

If only the operator was at fault, there are still several courses of corrective action. The operator could be cautioned, reinstructed, retrained, or removed from the job. It is usually not sufficient simply to caution an operator. Yet this is the action most frequently taken because it is easiest and it usually gets the person seeking corrective action off the back of the area supervisor. If this is not the best answer, what is? How can we achieve systematic corrective action so that the process of corrective action becomes institutionalized? To give the reader some idea of corrective action, Table 8.1 provides a sample list of types of corrective action and the clues that might lead to that action.

Organizing for Corrective Action

Although there are many ways that corrective action can be implemented, there is one method that is most effective. This method has proven time and again to be the most successful. All others either do not get the job done or fall apart in a relatively short time. What is this magic method?

1. *Form a team!* Call it a corrective action team, participative problem-solving team, or quality improvement team. The team must be empowered to introduce change. One member of the team must be given the leadership role and all team members must understand that their role is to analyze the process to determine how it can be improved. Then provide them with data and schedule regular meetings at the same time on the same day of the week at the same place at a prescribed interval. Generally, the interval should be weekly and rarely less frequent than weekly. Daily meetings are sometimes needed if production rates are high, but the meeting time should be correspondingly short and there should always be a weekly review of progress.

What are the ingredients for a successful corrective action program? The key ingredient is the people who attend. They should be the ones who are directly affected by the problems to be discussed. If production yields are the subject, always try to include manufacturing supervision, quality control or quality engineering, and industrial or manufacturing engineering. As necessary, people from other activities can be invited. Make sure that participants are those who understand how portions of the process are done so that analysis can be performed or that they have access to those who do understand the process.

2. *The meeting should be action oriented.* Do not waste time in great debates. Identify the problem, determine alternatives, and assign someone or a subgroup to investigate and report back at a subsequent time. Allow the person or group to set the schedule. Many problems are related to inadequate communications. These can be resolved at the

meeting. Often, someone attending the meeting will have a good idea about how to solve the problem. Try it!

3. *Issue minutes with progress status.* Include summaries of what happened at the meeting and make the report format cumulative so that a reader can see what has happened and what is planned. Then get the minutes issued promptly. As items are solved, drop them from the minutes.

4. *Come prepared to raise issues or answer action items.* If, after a few nonresponsive performances, action items are not addressed, do not invite that person again. Peer pressure usually keeps such action to a minimum.

5. *Make sure that something happens as a result of the meetings.* If it becomes a bull and gripe session and nothing happens, try to find a "spark plug" who will implement action. This is why it is important to invite people who are affected by the process to participate. These stakeholders are interested in making their own lives easier.

6. *Limit the meeting to an hour a week.* There may be rare occasions when it will last more than an hour, but do not let it become routine. If meetings are held daily, 15 minutes is all they should take. It may even be held on the floor with everyone standing during the session.

7. *Limit the number of items to be acted upon but allow time for new items to be brought up, discussed briefly, and either added to the list, solved, or dropped.* Do not limit the items to be brought up. If a problem is raised that lies within another area, invite others to participate or have the meeting chairman ask another group to address the problem.

8. *Focus on the problem, not the person.* Keep the meeting addressing problems constructively, not addressing people destructively.

9. *Use the data being gathered during the test and inspection operation as a source of problem information.* Also, use participants' experience to raise problems and suggest solutions. There are many times when it is important to solve problems outside the scope of the "quality data." Do not start with the toughest problem. Pick easy ones to begin with. Keep track of the difficult problems and apply steady pressure to the solutions. Do not get bogged down trying to cure the ills of the world and let the local, important problems go begging.

When Are Corrective Actions Developed?

The corrective action does not happen at the meeting. It happens between meetings. People are busy, very busy. However, corrective action

Table 8.1 Types of Corrective Actions Versus Clues

Types Of Corrective Action	*What Are The Clues?*
1. Change the design (specification)	a. Process capability analysis shows that the designed process can not meet requirements b. Competitive analysis shows that tighter requirements are not necessary c. The customer is not aware of the tighter specs and it does not produce a better product d. The customer or the next level of assembly does not require tighter requirements or the specification limits are shifted from what is needed.
2. Change the process	a. Defect levels remain undesirably high b. Capability of process is not there; market requires the tight spec; may have to buy equipment with better capability—if so, make sure of your spec needs; check for performance (calculate speed and feed, etc.)
3. Change the sequence	a. Flow analysis: defects are hidden by later operations; repairs are difficult; excessive defects
4. Change or repair a machine: machine differences, patterns, heads (filling)	a. Inspection or test results show one machine or one head from a machine producing higher defect rates. b. Plotting yields or other data in polar coordinates shows "bulges" in nonconformances
5. Invest in capital equipment	a. New equipment in market produces more consistently, smaller standard deviation to enable tighter tolerance or higher yields, and statistically-designed experiments have not reduced the product spread

Table 8.1 *Continued.*

Types Of Corrective Action	*What Are The Clues?*
6. Train operators and use their knowledge; statistical experimentation may be needed	a. Manual operations have high level of defects b. The suggestion system is ineffective c. Experienced operators are not given an opportunity to make suggestions d. Supervision came from the ranks and feels threatened or acts authoritarian e. Similar defects come from all operators
7. Overall cleanliness and environmental improvement needed	a. Product is sensitive to contaminants, and defects seem to occur randomly; look for open windows, irregular operation of other equipment (power surges), problems when certain operators work on product, hand or hair creams, lotions
8. Experiment with the process: include materials, locations, machines, operators, etc.; statistical experimentation needed	a. Difficult to reduce the spread of process; distribution of product parameters is not Gaussian (normal)
9. Switch vendors, correct vendors	a. One vendor's parts do not work as well and a significant difference exists among vendors; only one vendor is available and their products are erratic, have excessive standard deviations, or show multimodal distributions
10. Train supervision	a. Fear, wrong emphasis, lack of cooperation from body of work force
11. Change the operator	a. Everybody else is doing okay; one operator can not get it right: vision, dexterity, motivation
12. Change tools	a. Tolerances are exceeded, frequent tool changes necessary

Table 8.1 *Continued.*

Types Of Corrective Action	*What Are The Clues?*
13. Adjust setup	a. Process level is high/low or generally off-center
14. Change worst-case tolerancing to use statistical tolerances	a. Tolerances of parts are very tight and can not be held; assemblies are usually okay
15. Compensate for tool wear	a. Inspection or test results show gradual deterioration
16. Improve handling or storage protection	a. Electrostatic discharge can damage or degrade parts; parts show evidence of physical damage or degradation

is a necessity and it should be perceived as the way to make people productive in the long term. It should also improve job satisfaction. The meeting serves as the format for sharing information, exchanging ideas, and helping one another.

Follow-Up

This step is vital because implementation does not always happen as planned or the action recommended does not correct the problem. When action does not happen, support may be needed from higher levels of management or the task may simply need some extra effort. The corrective action chairman or leader should assign someone to follow up the action or do so himself or herself. If the solution gets implemented but does not correct the problem, a new analysis or corrective action is needed. The team should reinstitute the problem. The data being collected should provide insight to the effectiveness of the action.

It is essential to make sure that the problem disappears and stays solved. The result will be declining rejections and fewer problems. The effect of these events is multiplied. If, for example, rejects decline, inspection or test steps may be curtailed or even eliminated without product quality degradation. Rework, retest, reinspection, and scrap should decline with associated reduction in labor. If product quality improves, field performance should also improve creating a long-term benefit in terms of customer acceptance and eventual greater market penetration. In addition to the economic benefits, the work force will enjoy a better quality of work life when product quality is improved. Most people are proud of their work and resent slipshod approaches to manufacturing or service. Attention to corrective action is perceived as concern for the output—as indeed it is.

Conclusion

Corrective action is a critical ingredient in an effective quality system. Remember the four elements of corrective action: information, analysis, action and follow-up. Remember, too, the nine steps in the action cycle: people from several disciplines, regularly scheduled meetings, status reports, responsiveness, implementation, time-limited meetings, limited agenda, focus on problem, and use of the data. The results of such a program can be better products (or services), lower costs, greater market penetration, and greater personal satisfaction for employees. This approach is bound to bring forth positive results. Start it and keep it going. It will work for you.

9 · Fundamentals of Sampling

Introduction

In this chapter we discuss statistical sampling. Sampling has come under attack recently by Deming (1987) and others who perceive it as the wrong way to control quality and, worse, to lull the user into a false sense of security that the product, process, or service is adequate when it really needs improvement. The better way to control quality, they argue, is through process control and continuous improvement. Although this is true, there are instances when statistical sampling is useful. Sampling plans are also used to determine the acceptability of products, processes, and services, such as the accuracy of accounting and auditing procedures, on-time arrivals of airplanes, cleanliness of hotel rooms, advertising billing accuracy, and other administrative processes.

Proper understanding and use of statistical sampling procedures can result in significant cost savings as well as product, process, or service improvement. To obtain maximum benefit from this tool of statistical quality control it is necessary to be aware of the types of sampling plans available, their limitations, and the possible areas of application.

In this chapter we introduce some of the sampling plans available and describe the most commonly used procedures.

Sampling procedures are used to determine what materials or processes have a sufficiently high percentage of conforming items to be suitable for use. Ideally, the materials or processes will contain no nonconforming items. Therefore, no sampling is needed to determine compliance. This will be true if a product is manufactured under a state of statistical control and the process capability is better than specifications require. If, on the other hand, the process is not capable of producing within specification limits, or is not in statistical control, some form of inspection is in order. The problem is, when confronted with a lot of materials of unknown origin or with an unknown process, to determine whether or not there is conformance.

Statistical sampling procedures offer an economical way of making the determination with controlled risks based on rules of probability. Even if product is 100% inspected or tested, it is desirable to follow up with sampling because of the inability of 100% inspection or test to remove all nonconformities. Sampling provides insight to the product quality and, when records of results are maintained, will provide information about the nature of the problem and may lead to corrective action.

Sampling plans fall into two general categories, attributes and variables, with occasional plans combining both. Attributes are characteristics determined qualitatively, either acceptable or not acceptable. Variables are characteristics that are determined quantitatively and therefore have a measurement. In addition, sampling plans can be based on single-sampling, double-sampling, or multiple- or sequential sampling plans. These plans enable an accept or reject decision after one, two, or many samples. There are also other types of plans, such as continuous sampling, multilevel continuous sampling, chain sampling, skip-lot sampling, and narrow limit sampling. For more details on sampling, see Shilling (1982).

The most common form of acceptance sampling uses attributes and is documented in MIL-Std 105D or ANSI Standard Z1.4. When using single sampling, the lot disposition is determined after the first sample has been completed. In double sampling, a second sample may be taken with the decision to accept, reject, or evaluate the second sample after the first sample has been completed. The second sample then enables an accept or reject decision.

Although in double and multiple sampling fewer samples are taken on the average than in single sampling, the probabilities of acceptance are the same as for corresponding single-sampling plans. Additional training and administration are required with double sampling and still more for multiple sampling. In multiple sampling, this process is iterated seven times with an accept, reject, or take-the-next-sample decision after all but the final sample. In this document, sampling plans are classified by the acceptable quality level (AQL) or limiting quality (LQ). Other attribute plans, such as those in Dodge and Romig (1959), are classified by average outgoing quality limit (AOQL) or lot tolerance percent defective (LTPD). These terms are defined later.

Variables sampling requires that measurements be taken. It utilizes the concept of the average and standard deviation (or range) to determine the degree of product conformance. Its biggest advantage is that many fewer pieces are required for a decision. Its biggest disadvantage is that its use requires a distribution for the measured characteristic to be known and have a normal (bell-shaped) distribution or be close to normal to be valid. It also requires a measurement and some calculations. With automated equipment, calculators and computers, this is no longer a problem. As a result, the use of variables sampling plans is increasing.

Continuous sampling is used most often to control a process. It is operated by performing 100% inspection until a predetermined number of consecutively good pieces are found, then taking a percentage of subsequent products. In multilevel continuous sampling, the percentage is adjusted in steps based on results. Interestingly, in attribute and variables sampling, taking samples that are percentages of a lot is absolutely the incorrect thing to do since that has a major affect on probabilities of acceptance (see the section "The Fallacy of 10 Percentitis").

Operating Characteristic Curve

An operating characteristic (OC) curve is used to show the probabilities associated with a sampling plan. This curve (Fig. 9.1) is a plot of the quality of submitted lots or processes in percentage defective (p) versus its probability of acceptance (P_a). It is a pictorial representation of the ability of the plan to discriminate between good and bad lots.

There are two types of error associated with statistical sampling. First is the error of rejecting a good lot because, occasionally, the sample will contain more nonconforming items than the plan allows. This is known as the *producer's risk* or, alternatively, as *type I error* or *alpha error.* The value of this risk can be measured and is in the region of the AQL. There is also the probability of accepting a lot that is worse

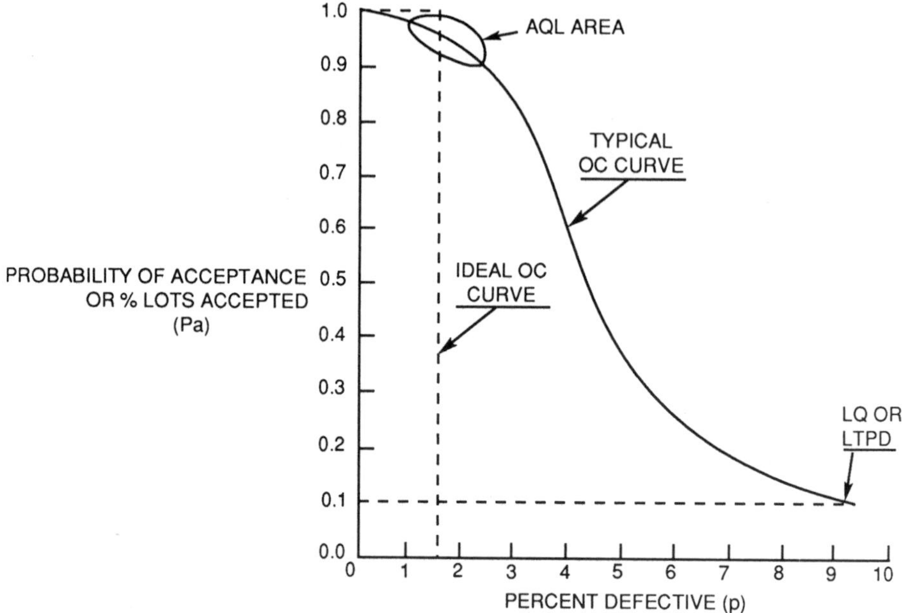

Fig. 9.1 Operating characteristic curve.

than desired because, occasionally, the sample contains fewer nonconformances than the plan allows. This is called the *consumer's risk*, or, alternatively, *type II error* or *beta error.* This is usually at the point of the LQ for 10% or the LTPD.

The slope of the OC curve determines the ability of the sampling plan to discriminate between acceptable and unacceptable lots. The steeper the slope, the better the discrimination. The ideal plan would be a horizontal line extending from the 100% P_a to the predetermined acceptable level and then a vertical line extending to the 0% P_a. This is shown as the dashed curve in Fig. 9.1. This ideal curve cannot even be achieved by 100% inspection unless the process is both capable and under statistical control. That is why process control is the only way to produce the defect-free product required in achieving world-class quality.

Some of the significant points on the operating characteristic curve are shown in Fig. 9.1. These include:

1. The acceptable quality level (AQL) region
2. The limiting quality (LQ) of lot tolerance percent defective (LTPD)

The acceptable quality level (AQL) is the maximum percent defective (or the maximum number of defects per hundred units) that for purposes of sampling inspection can be considered satisfactory as a process average. The process average, which is the best measure of the quality of the lot submitted, is determined by dividing the cumulative number of defects found in the samples by the cumulative number of items inspected. Some texts recommend that 10 lots be used and that only first-sample results be accumulated. Other texts recommend that all sample results be used. The author favors first-sample results only to avoid biasing the data in favor of poor lots (because larger cumulative sample sizes will result from poorer-quality lots). The process average should be lower in value (perhaps 10%) than the AQL.

The limiting quality (LQ) is defined as the percent defective at a given probability of acceptance such as 10 or 5%. The lot tolerance percent defective (LTPD) is identical for an LQ for which the probability of acceptance is 10%. These plans are most often used for occasional lots or batches (rather than a continuous series of lots or batches) or where the consumer's risk is critical. These types of plans offer protection to the consumer that lots will exceed the designated LQ value only 10 or 5% of the time, depending on the plan selected.

There are two additional terms that cannot be represented directly on the OC curve: the average outgoing quality limit (AOQL) and the discrimination ratio. The AOQL is shown on the average outgoing quality (AOQ) curve in Fig. 9.2. It is derived by multiplying the percent nonconforming, p, by the probability of acceptance, P_a. and plotting the product against p. It is the maximum value of the average outgoing quality.

The *average outgoing quality* (AOQ) is defined as the average quality of the outgoing product, including all accepted lots (which may

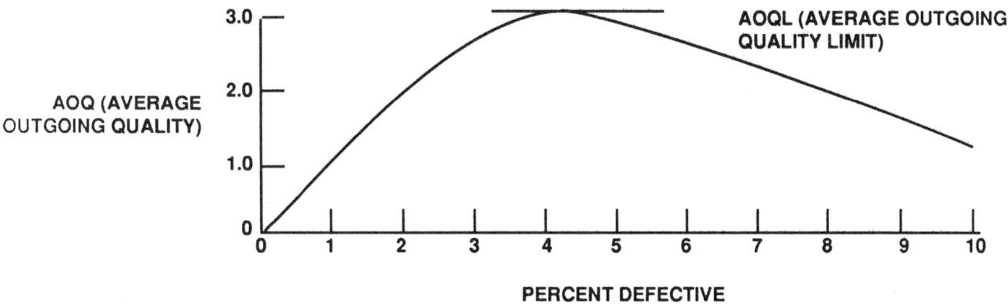

Fig. 9.2 Average outgoing quality curve.

contain nonconformances in the uninspected portion), plus all rejected lots after the rejected lots have been effectively 100% inspected and all the nonconforming items replaced by conforming items. The *average outgoing quality limit* (AOQL) is the maximum of the AOQs for all possible incoming qualities for a given acceptance sampling plan. Therefore, it assumes that all rejected lots are not only defect free but that the product stream contains these pure lots. The *discrimination ratio* is the quotient of the percent nonconforming values at P_a = 10% probability of acceptance and at P_a = 90% probability of acceptance on the OC curve.

Advantages and Disadvantages of Sampling Plans

In a world-class environment we must recognize that quality levels must be very high and costs must be very low. Anytime that inspection or test is used, whether 100% or sampling, costs are increased. It should be a constant and ongoing objective to eliminate these costs while maintaining processes that produce conforming products. What is really needed is a process whose capability (variation from highest to lowest value or from the best to the worst) is much smaller than the specification limits or better than customer requirements. As more companies strive for inventory reductions and more inventory turnovers, the concept of just-in-time (JIT) delivery will continue to expand. The JIT concept virtually demands that products be good enough to go from the receiving dock directly into the production line. There is no time for inspection by any method.

In a similar fashion, in-process inspection and testing create added costs that should be eliminated when processes are good enough to produce defect-free products. Emphasis must be placed on process control to assure full product conformance. Why, then, must we address statistical sampling? Simply because many companies have not arrived at the quality levels necessary to compete for world-class quality. So an understanding of what is available and, in general, how to put what is available into constructive application is necessary.

Advantages

1. There can be a significant reduction in costs with no significant decrease in quality; sampling inspection is less costly than 100% inspection provided that the product or process being sampled (whose

quality level is unknown) is at the level needed by the user. The supplier is motivated to produce a better quality product because they run the risk of having entire lots rejected rather than just defective items. Moreover, by requiring 100% lot reinspection by the producer when a lot is rejected, the producer will be even more careful about the quality of the products submitted. This is enforceable when used in-house but may be difficult when dealing with suppliers and should therefore be a part of the purchase order.

2. A necessary part of a sampling plan is the requirement to tighten (increase the severity of) inspection when certain reject levels are exceeded. Such tightening generally prohibits prolonged operation at poor levels of quality. This creates further pressure to maintain consistently high quality to pass acceptance sampling. These procedures should not be applied when lot rejections are 20 to 25% of the lots submitted. Then correction is required or 100% inspection must precede sampling. ANSI Standard Z1.4 provides for discontinuance of sampling if 10 consecutive lots are on tightened inspection.

3. Sampling provides a level of assurance not present when only 100% inspection is performed. It has long been recognized that 100% inspection does not identify all the defective units. This may be due to operator or equipment error, material nonrepeatability or change, or other, unknown causes. Although this condition is changing with the use of automated inspection and testing, one must be aware of the nature of the 100% inspection operations to make an intelligent decision. Operator errors due to inattention or fatigue may be considerably reduced by sampling inspection because each sample is considered a discernible entity rather than an endless flow of material.

4. Sampling provides insight into product quality and, where records of results are maintained, will provide information about the nature of the problem and may lead to corrective action. Of course, this is true of 100% inspection as well, but there are times when too many data obscure facts, and sampling results can provide quick insight to problems and possible causes. In their book *Human Factors in Quality Assurance,* Harris and Chaney (1969) conducted a study of inspection efficiency. The efficiency ranged between 10 and 90% depending on product complexity. That means that between 1 and 9 defects are found for every 10 that are present. As noted, advances in inspection technology may have improved these numbers, but we cannot assume that 100% inspection or testing removes all defects. Once again, the key is to have the process capable and under control.

5. Some materials are destroyed during evaluation. This requires sampling techniques for evaluation of those quality characteristics that cannot be evaluated without destruction of the product.

6. The evaluation may be so expensive that some form of sampling is necessary for economic reasons.

Disadvantages

Problems created by using sampling plans are generally minor in nature and are more than offset by the gains to be realized.

1. Training must be performed to acquaint inspection, manufacturing, and supervisory personnel in the use of sampling.

2. Suppliers may be unwilling to accept lot rejections based on sampling, and an agreement has to be reached and incorporated in the purchase order.

3. Occasional good lots will be rejected. This is the producer's risk (sometimes called alpha and type I error) and will result in added costs due to unnecessary reinspection. Any attribute sampling plan for which the acceptance number is zero has an increased level of producer's risk. If a supplier's process is operating at or below the value of the AQL, the large majority of lots (about 95% and higher) will be accepted. So the pressure for improvement does not exist under these circumstances unless there is recognition that continued prosperity is a function of continued improvement.

4. Occasional bad lots will be accepted, creating problems. This is termed consumer's risk (sometimes called beta or type II error).

Items 3 and 4 are risks associated with all sampling and must be recognized.

Application of Sampling Procedures

Wherever inspection or testing takes place, sampling may be used. This includes evaluation of product upon receipt, in-process inspection as items flow through a process, and final inspection before delivery and in the field. Some other precautions that must be observed in applying sampling techniques are worthy of note.

1. Sampling alone cannot be used effectively to upgrade the quality of products from a controlled process. To upgrade product quality, some definitive change must be introduced in the process.

2. The lot being sampled should be homogeneous within the limits of practicality. If, for example, material from a poor-quality-level production line is intermingled with material from a high-quality-level production line, the sampling plan will merely reject some lots and accept others based on the laws of probability. If, however, these high-quality and low-quality lots are not mixed, the sampling plan will reject a much higher percentage of the poor-quality lots. It may, indeed, identify the source of poor quality and point the way to corrective action. So the location of the sampling operation in the product flow is important.

3. Another precaution to be observed is the method of selecting the sample. The textbooks state that samples should be selected at random. There are tables of random numbers available that can be used. However, the idea behind random sampling is to obtain a representative sample lacking knowledge of the distribution of nonconforming items in the lot. The representation is what is important. Thus each item in the lot should have an equal chance of being selected. At times, it is difficult to obtain a representative sample. For example, how is such a sample obtained from spools of wire? Is it satisfactory to test pieces selected only from the ends? It may be acceptable to do this provided that uniformity throughout spools has been determined. This may be done by unspooling and testing or by using wire and finding no difficulties throughout its length. Each situation has to be evaluated individually. Generally, samples are selected through attempts of randomization. These are often sufficient. Just do not determine the acceptability of a basket of fruit by looking only at the top layer.

4. It is desirable to maintain accurate records to enable the process average calculation and to identify types of defects found. (The process average is the percentage of nonconforming items in the cumulative first-sample results. Often this accumulation is limited to a moving window of 10 lots.) As a minimum, date, lot size, sample size, number of defects found, the types of defects found, and lot disposition must be kept. The lot identity should also be recorded to facilitate corrective action. See Fig. 9.3 for a suitable form for sampling inspection. Note that nonconformance categories are in columns to facilitate analysis. The column with the most recordings or highest quantity shows the largest number of nonconformances and therefore the greatest opportunity for improvement.

5. Be aware of inspector bias. When defects found are too consistently at the acceptance number, it is time to check inspector performance. Some inspectors are inclined to avoid rejection and so avoid

DATE	LOT SIZE	FIRST SAMPLE				SECOND SAMPLE			LOT ACTION		DETAIL DEFECTS							
		NO. INSP	NO. REJ	NO. ALLOW	% DEF	NO. INSP	NO. REJ	NO. ALLOW	A C	R E								

Fig. 9.3 Typical form for acceptance sampling by attributes: double sampling.

finding the last defect that will reject a lot. Others seek to have the highest reject rate and will stretch a point to find the last defect and thus reject the lot. If the process average is maintained, comparing the percentage of lots accepted for that process average, p, with the operating characteristic curve can provide a check. It should approximate the value of P_a where p intersects the curve. This could be a nuisance, but the computer can do it automatically.

6. Switching between single- and double-sampling plans during a single-lot evaluation is improper because OC curves are based on probability and probabilities are different when this is done. The degree of protection or discrimination afforded by the sampling plan will therefore change if the sampling rules are not followed.

7. Do not use a fixed percentage of the lot as a sample size because the OC curve changes drastically as the lot size changes when this is done.

The Fallacy of 10 Percentitis

There are times when it is tempting simply to state that a useful sampling plan is to take a fixed percentage of the lot as the sample size and allow no defects (or some small number of defects) for a quick and easy sampling plan. For example, take 10% and allow no defects. This is wrong if lot sizes vary. One of the objectives of sampling is to use plans that provide known levels of protection. When a plan using a sample size of 10% of the lot and 0 defects allowable for acceptance purposes, the degree of protection varies widely because the OC curve is very sensitive to sample size (Fig. 9.4). When in doubt, therefore, it is best to use a fixed sample size regardless of the lot size, because the OC curves are not greatly affected by lot size differences. As you shall see in the balance of this chapter, sampling plans are based on probabilities, and these must be taken into consideration when deviating from published tables.

Types of Sampling Plans

In determining what types of sampling plans to use, we must be aware of the types that are available—principally attributes and variables plans.

Attributes Sampling

The most popular sampling plans are attribute plans and those in ANSI Standard Z1.4 (MIL-St 105D) are most often used (see Appendix A at the end of the chapter). In these documents, sampling plans are

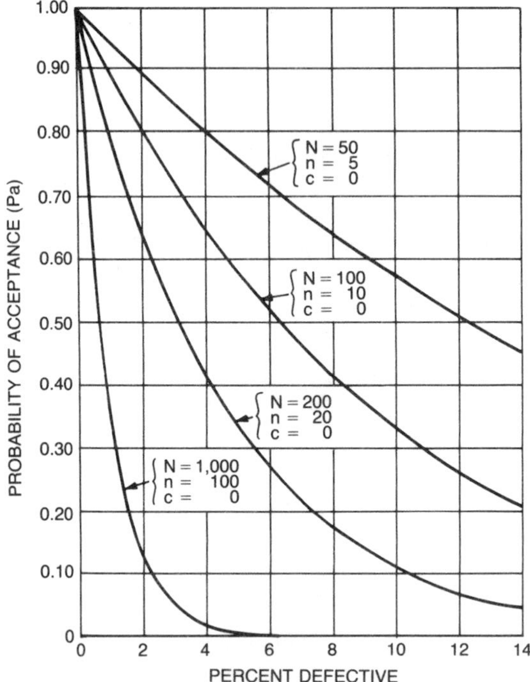

Fig. 9.4 Comparison of operating characteristic curves for four different sampling plans involving 10% sampling.

catalogued by AQL (acceptable quality level), which is defined as the maximum percent defective that for the purposes of sampling inspection can be considered as a process average. There is also a classification by limiting quality (LQ) at 5 and 10%; the limiting quality is that percent nonconforming that has a 5 or 10% probability of acceptance, respectively.

The ANSI Standard plans are easier to apply than Dodge-Romig when more than one level of inspection is chosen. Thus critical characteristics can be sampled at a 0.4% AQL, major characteristics at 2.5% ALQ, and minor characteristics at 6.5% AQL, and all of these will usually allow the same sample size but different acceptance numbers. This simplifies sample selection, record keeping, and inspection procedures.

Classifying plans by AQL results in the operating characteristics (OC) curves being close together in the region of the AQL—that region where there is a high probability of acceptance. However, in the region of OC curves where there is a low probability of acceptance, the curves may vary widely, with smaller sample sizes (corresponding to smaller lots) having a much higher probability of accepting poorer-quality lots. Thus consumer protection is lower for smaller lots. To prevent this, a minimum-sample-size code letter (or lot size) is sometimes used. Alternatively, a maximum LQ (or LTPD) may be specified.

On the other hand, the Dodge–Romig tables for AOQL are set up for minimum sampling. The costs associated with this type of sampling are thus minimal. Whereas the AQL can be found on the OC curve, the AOQL has no meaning in relation to the OC curve. The percentage value for AOQL usually lies in the 40 to 50% probability of acceptance range on the OC curve. As such, a 1% AQL plan is always looser than a 1% AOQL plan (Fig. 9.5) and a 1% LTPD plan is tighter still. The AOQL concept was established for the Bell Telephone System at a time when all products were delivered to the same in-house customer.

An important feature of AOQL plans is that the AOQ value is based on 100% screening of rejected lots, replacement of all defects with

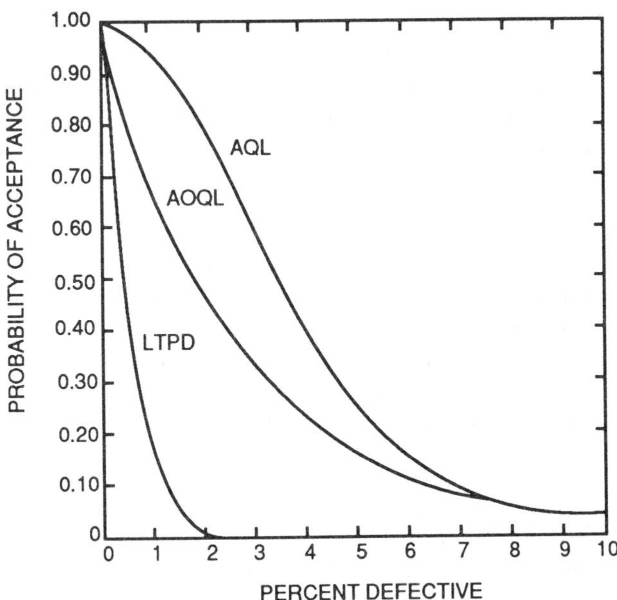

Fig. 9.5 Comparison of sperating characteristic curves for 1% AQL, 1% AOQL, and 1% LTPD.

good units. However, a 100% inspection will usually not remove all the defects, so the true AOQL might be slightly higher than stated in the tables. A more important consideration when there are multiple customers is that the lot that fails may be diverted from production (rather than 100% inspected) and then shipped to a customer with less rigorous requirements. In this case the actual AOQL value may be considerably higher. The Dodge–Romig tables for lot tolerance percent defective (LTPD) are normally specified when isolated lots are submitted or when consumer protection is crucial. This determination is frequently based on product type and usage. ANSI Standard Z1.4 uses LQ at 10% P_a instead of LTPD, but the terms are identical. Standard Z1.4 also has an LQ at $P_a = 5\%$.

Variables Sampling

In variables sampling, the characteristics being evaluated must have actual measurements associated with them. In addition, the shape of the distribution from which the sample has been selected must be known. Although this is a severe limitation, variables sampling plans offer significant reduction in sample sizes. The more common of these sampling plans are those utilizing a normal or Gaussian distribution. Variables plans are available for known standard deviations and unknown standard deviation for both single- and double-sided specifications. The most often used set of plans are in ANSI Standard Z1.9, which is also MIL-Std 414.

Summary

I have tried to provide a brief overview of sampling: its uses, limitations, and advantages. The fact that sampling plans do exist, are used, and contribute to running a business makes it useful to be aware of how these plans are used. There are many other applications to be sure, but for the purposes of this book, enough information has been provided. See Appendix A for a specific application of sampling procedures.

Appendix A: Using Sampling Plans

ANSI Z1.4 (Military Standard 105D) Tables

To provide the reader with a brief review of how sampling plans are used, the following discussion is presented relative to ANSI Z1.4. First, the inspection level and the lot size must be known; Table 9.1 (Table I of MIL-Std 105D) is used for this purpose. General inspection level II is always the starting point unless otherwise specified. Level I may be used when less discrimination is needed, and level III may be used for greater discrimination; however, these inspection levels should not be confused with reduced or tightened inspection, which will be discussed later. The four levels, Sl, S2, S3, and S4, are used when small sample sizes are necessary and large sampling risks can or must be tolerated, as would be the case where samples are destroyed in testing or test costs are very high. Using the left-hand column for lot or batch size, a determination can then be made as to the inspection letter to be used. For example, suppose that a lot size of 1000 pieces is to be evaluated. From Table 9.1, using general inspection level II, code letter J is selected as the proper sample size letter. This inspection level actually determines the relationship between the sample size and the lot size and is used in conjunction with the AQL value to select a single-, double-, or multiple-sampling plan.

Table 9.1 Sample Size Code Letters

	Special inspection levels				*General inspection levels*		
Lot or batch size	*S-1*	*S-2*	*S-3*	*S-4*	*I*	*II*	*III*
2– 8	A	A	A	A	A	A	B
9– 15	A	A	A	A	A	B	C
16– 25	A	A	B	B	B	C	D
26– 50	A	B	B	C	C	D	E
51– 90	B	B	C	C	C	E	F
91– 150	B	B	C	D	D	F	G
151– 280	B	C	D	E	E	G	H
281– 500	B	C	D	E	F	H	J
501– 1,200	C	C	E	F	G	J	K
1,201– 3,200	C	D	E	G	H	K	L
3,201– 10,000	C	D	F	G	J	L	M
10,001– 35,000	C	D	F	H	K	M	N
35,001–150,000	D	E	G	J	K	N	P
500,001 and over	D	E	H	K	N	Q	R

Source: ANSI/ASQC Z1.4–1981, *Sampling Procedures and Tables for Inspection by Attributes*, American Society for Quality Control, Milwaukee, Wis., 1981

Single Sampling

Having now determined the inspection level and assuming that single sampling is used, Table 9.2 (Table II-A of MIL-Std 105D), the master table for single sampling, is used. This is the most straightforward and easily administered sampling procedure. Table II-A shows that for code letter J, the sample size should be 80 pieces. The AQL must then be decided upon so that the allowable number of nonconformities may be determined. Assuming an AQL of 1.0%, the acceptance number is 2 and the reject number is 3. If in a sample of 80 pieces, two or fewer nonconformities are found, the entire lot may be accepted; if three or more are found, the lot must be rejected.

One of the problems to be solved in using sampling plans is the determination of the value of the AQL. This may actually be calculated by taking the ratio of the cost of inspection to the cost of processing a nonconforming item at the next inspection point. This assumes that the nonconformance will be detected. If it is not, the cost associated with further processing or actual field failures enters the equation, and the calculation is much more complex. Frequently, however, these cost figures are not known. As a result, the method most often used is to select the AQL value based on the critical nature of the nonconformance, its effect on future assembly or use, and the manufacturing method used to produce the item. Note that the lowest level of AQL, 0.010, is equivalent to a 100-ppm level of nonconformance. The AQL level should not be confused with the process average, which estimates the lot quality. Earlier in this chapter, the statement was made that the process average should be considerably less than the AQL value and 10% was suggested. Thus a process operating at a 10-ppm level is needed to pass this sampling plan consistently.

Some further explanation in the use of the table is necessary. When a sample size and AQL combine to intersect at an arrow in the table, the direction of the arrow should be followed to the first available sampling plan, either above or below. Both the sample size and the accept and reject numbers comprise the sampling plan and both must be used at the same time. One frequent error is the use of only the accept or reject number above the arrow in conjunction with the original sample size. For example, if the AQL were specified at 0.65, the sampling plan would be 125 pieces and 1 nonconformity allowed. If 25% AQL (or defects per hundred) were specified, the plan would be take 50 pieces and allow 18 nonconformities (or 18 defects per hundred).

Table 9.2 Single-Sampling Plans for Normal Inspection (Master Table)

Sample size code letter	Sample size	Acceptable Quality Levels (normal inspection) 0.010	0.015	0.025	0.040	0.065	0.10	0.15	0.25	0.40	0.65	1.0	1.5	2.5	4.0	6.5	10	15	25	40	65	100	150	250	400	650	1000
		Ac Re	Ac Re	Ac Re	Ac Re	Ac Re	Ac Re	Ac Re	Ac Re	Ac Re	Ac Re	Ac Re	Ac Re	Ac Re	Ac Re	Ac Re	Ac Re	Ac Re	Ac Re	Ac Re	Ac Re	Ac Re	Ac Re	Ac Re	Ac Re	Ac Re	Ac Re
A	2	↓	↓	↓	↓	↓	↓	↓	↓	↓	↓	↓	↓	↓	↓	0 1	↓	↓	1 2	2 3	3 4	5 6	7 8	10 11	14 15	21 22	30 31
B	3	↓	↓	↓	↓	↓	↓	↓	↓	↓	↓	↓	↓	↓	0 1	↑	↓	1 2	2 3	3 4	5 6	7 8	10 11	14 15	21 22	30 31	44 45
C	5	↓	↓	↓	↓	↓	↓	↓	↓	↓	↓	↓	↓	0 1	↑	↓	1 2	2 3	3 4	5 6	7 8	10 11	14 15	21 22	30 31	44 45	↑
D	8	↓	↓	↓	↓	↓	↓	↓	↓	↓	↓	↓	0 1	↑	↓	1 2	2 3	3 4	5 6	7 8	10 11	14 15	21 22	30 31	44 45	↑	↑
E	13	↓	↓	↓	↓	↓	↓	↓	↓	↓	↓	0 1	↑	↓	1 2	2 3	3 4	5 6	7 8	10 11	14 15	21 22	30 31	44 45	↑	↑	↑
F	20	↓	↓	↓	↓	↓	↓	↓	↓	↓	0 1	↑	↓	1 2	2 3	3 4	5 6	7 8	10 11	14 15	21 22	↑	↑	↑	↑	↑	↑
G	32	↓	↓	↓	↓	↓	↓	↓	↓	0 1	↑	↓	1 2	2 3	3 4	5 6	7 8	10 11	14 15	21 22	↑	↑	↑	↑	↑	↑	↑
H	50	↓	↓	↓	↓	↓	↓	↓	0 1	↑	↓	1 2	2 3	3 4	5 6	7 8	10 11	14 15	21 22	↑	↑	↑	↑	↑	↑	↑	↑
J	80	↓	↓	↓	↓	↓	↓	0 1	↑	↓	1 2	2 3	3 4	5 6	7 8	10 11	14 15	21 22	↑	↑	↑	↑	↑	↑	↑	↑	↑
K	125	↓	↓	↓	↓	↓	0 1	↑	↓	1 2	2 3	3 4	5 6	7 8	10 11	14 15	21 22	↑	↑	↑	↑	↑	↑	↑	↑	↑	↑
L	200	↓	↓	↓	↓	0 1	↑	↓	1 2	2 3	3 4	5 6	7 8	10 11	14 15	21 22	↑	↑	↑	↑	↑	↑	↑	↑	↑	↑	↑
M	315	↓	↓	↓	0 1	↑	↓	1 2	2 3	3 4	5 6	7 8	10 11	14 15	21 22	↑	↑	↑	↑	↑	↑	↑	↑	↑	↑	↑	↑
N	500	↓	↓	0 1	↑	↓	1 2	2 3	3 4	5 6	7 8	10 11	14 15	21 22	↑	↑	↑	↑	↑	↑	↑	↑	↑	↑	↑	↑	↑
P	800	↓	0 1	↑	↓	1 2	2 3	3 4	5 6	7 8	10 11	14 15	21 22	↑	↑	↑	↑	↑	↑	↑	↑	↑	↑	↑	↑	↑	↑
Q	1250	0 1	↑	↓	1 2	2 3	3 4	5 6	7 8	10 11	14 15	21 22	↑	↑	↑	↑	↑	↑	↑	↑	↑	↑	↑	↑	↑	↑	↑
R	2000	↑	↑	1 2	2 3	3 4	5 6	7 8	10 11	14 15	21 22	↑	↑	↑	↑	↑	↑	↑	↑	↑	↑	↑	↑	↑	↑	↑	↑

↓ = Use first sampling plan below arrow. If sample size equals, or exceeds, lot or batch size, do 100 percent inspection.
↑ = Use first sampling plan above arrow.
Ac = Acceptance number.
Re = Rejection number.

Source: ANSI/ASQC Z1.4–1981, *Sampling Procedures and Tables for Inspection by Attributes*, American Society for Quality Control, Milwaukee, Wis., 1981

One of the advantages in the use of sampling plans is that inspection is adjusted automatically based on product quality. If the products are exceedingly poor, sampling plans will require more inspection because more lots will be rejected and thus will need 100% inspection to purge them of nonconformances. The sampling itself will average fewer items inspected because lots will be rejected sooner if sampling is terminated when the rejection number is reached. In the ANSI or Military Standard tables, this is achieved by requiring the use of tightened inspection. See Table 9.3 (Table II-B of MIL-Std 105D) for the tightened inspection tables. When two lots out of five consecutive lots have been rejected on original inspection, tightened inspection is imposed automatically. If this were to occur in the sampling plan in a lot of 1000, it would mean that our sample size would remain the same but that the allowable number of nonconformances would be reduced from two to one. Tightened inspection is generally one AQL level tighter than normal inspection. Thus the plan for a 0.65% AQL normal is the same as for a 1% AQL tightened. When on tightened inspection, normal inspection may be reinstated when five consecutive lots pass inspection on original submission.

Just as there are rules for going to tightened inspection, rules exist for going to reduced inspection. These rules require that the preceding 10 lots have been on normal inspection and none has been rejected, that the total nonconformities in the preceding 10 lots are equal to or less than the tabulated value in Table 9.4 (Table VIII of MIL-Std 105D), that production is at a steady rate, and that reduced inspection is considered desirable by the responsible authority. When reduced inspection is allowed, the sampling plan used is determined from Table 9.5 (Table II-C of MIL-Std 105D) for single sampling.

If reduced inspection is in effect, the sampling plan for the lot of 1000 at an AQL of 1.0% is sample size 32: Accept if one or less nonconformance is found and reject if three or more nonconformances are found; if two nonconformances are found, the lot is accepted but normal inspection is reinstated. In general, the reduced sample size is 40% of the single-sample size. Normal inspection is also reinstated when a lot is rejected or production becomes irregular or "other conditions warrant that normal inspection shall be instituted."

Since reduced inspection is optimal, the purchaser or consumer may at his or her discretion terminate reduced inspection. This is different from tightened inspection in that tightened inspection is a mandatory requirement for most sampling plans.

Table 9.3 Single-Sampling Plans for Tightened Inspection (Master Table)

Sample size code letter	Sample size	Acceptable Quality Levels (tightened inspection)																									
		0.010	0.015	0.025	0.040	0.065	0.10	0.15	0.25	0.40	0.65	1.0	1.5	2.5	4.0	6.5	10	15	25	40	65	100	150	250	400	650	1000
		Ac Re	Ac Re	Ac Re	Ac Re	Ac Re	Ac Re	Ac Re	Ac Re	Ac Re	Ac Re	Ac Re	Ac Re	Ac Re	Ac Re	Ac Re	Ac Re	Ac Re	Ac Re	Ac Re	Ac Re	Ac Re	Ac Re	Ac Re	Ac Re	Ac Re	Ac Re
A	2	↓	↓	↓	↓	↓	↓	↓	↓	↓	↓	↓	↓	↓	↓	↓	↓	↓	↓	1 2	2 3	3 4	5 6	8 9	12 13	18 19	27 28
B	3	↓	↓	↓	↓	↓	↓	↓	↓	↓	↓	↓	↓	↓	↓	0 1	↓	↓	1 2	2 3	3 4	5 6	8 9	12 13	18 19	27 28	41 42
C	5	↓	↓	↓	↓	↓	↓	↓	↓	↓	↓	↓	↓	↓	0 1	↓	↓	1 2	2 3	3 4	5 6	8 9	12 13	18 19	27 28	41 42	↑
D	8	↓	↓	↓	↓	↓	↓	↓	↓	↓	↓	↓	↓	0 1	↓	↓	1 2	2 3	3 4	5 6	8 9	12 13	18 19	27 28	41 42	↑	↑
E	13	↓	↓	↓	↓	↓	↓	↓	↓	↓	↓	↓	0 1	↓	↓	1 2	2 3	3 4	5 6	8 9	12 13	18 19	27 28	41 42	↑	↑	↑
F	20	↓	↓	↓	↓	↓	↓	↓	↓	↓	↓	0 1	↓	↓	1 2	2 3	3 4	5 6	8 9	12 13	18 19	↑	↑	↑	↑	↑	↑
G	32	↓	↓	↓	↓	↓	↓	↓	↓	↓	0 1	↓	↓	1 2	2 3	3 4	5 6	8 9	12 13	18 19	↑	↑	↑	↑	↑	↑	↑
H	50	↓	↓	↓	↓	↓	↓	↓	↓	0 1	↓	↓	1 2	2 3	3 4	5 6	8 9	12 13	18 19	↑	↑	↑	↑	↑	↑	↑	↑
J	80	↓	↓	↓	↓	↓	↓	↓	0 1	↓	↓	1 2	2 3	3 4	5 6	8 9	12 13	18 19	↑	↑	↑	↑	↑	↑	↑	↑	↑
K	125	↓	↓	↓	↓	↓	↓	0 1	↓	↓	1 2	2 3	3 4	5 6	8 9	12 13	18 19	↑	↑	↑	↑	↑	↑	↑	↑	↑	↑
L	200	↓	↓	↓	↓	↓	0 1	↓	↓	1 2	2 3	3 4	5 6	8 9	12 13	18 19	↑	↑	↑	↑	↑	↑	↑	↑	↑	↑	↑
M	315	↓	↓	↓	↓	0 1	↓	↓	1 2	2 3	3 4	5 6	8 9	12 13	18 19	↑	↑	↑	↑	↑	↑	↑	↑	↑	↑	↑	↑
N	500	↓	↓	↓	0 1	↓	↓	1 2	2 3	3 4	5 6	8 9	12 13	18 19	↑	↑	↑	↑	↑	↑	↑	↑	↑	↑	↑	↑	↑
P	800	↓	↓	0 1	↓	↓	1 2	2 3	3 4	5 6	8 9	12 13	18 19	↑	↑	↑	↑	↑	↑	↑	↑	↑	↑	↑	↑	↑	↑
Q	1250	↓	0 1	↓	↓	1 2	2 3	3 4	5 6	8 9	12 13	18 19	↑	↑	↑	↑	↑	↑	↑	↑	↑	↑	↑	↑	↑	↑	↑
R	2000	0 1	↑	↓	1 2	2 3	3 4	5 6	8 9	12 13	18 19	↑	↑	↑	↑	↑	↑	↑	↑	↑	↑	↑	↑	↑	↑	↑	↑
S	3150			1 2																							

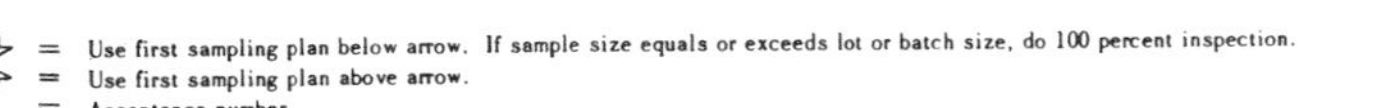

↓ = Use first sampling plan below arrow. If sample size equals or exceeds lot or batch size, do 100 percent inspection.
↑ = Use first sampling plan above arrow.
Ac = Acceptance number.
Re = Rejection number.

Source: ANSI/ASQC Z1.4–1981, *Sampling Procedures and Tables for Inspection by Attributes*, American Society for Quality Control, Milwaukee, Wis.

Table 9.4 Limit Numbers for Reduced Inspection

Number of sample units from last 10 lots or batches	Acceptable Quality Level																								
	0.010	0.015	0.025	0.040	0.065	0.10	0.15	0.25	0.40	0.65	1.0	1.5	2.5	4.0	6.5	10	15	25	40	65	100	150	250	400	650	1000
20-29	•	•	•	•	•	•	•	•	•	•	•	•	•	•	•	0	0	2	4	8	14	22	40	68	115	181
30-49	•	•	•	•	•	•	•	•	•	•	•	•	•	•	0	0	1	3	7	13	22	36	63	105	178	277
50-79	•	•	•	•	•	•	•	•	•	•	•	•	•	0	0	2	3	7	14	25	40	63	110	181	301	
80-129	•	•	•	•	•	•	•	•	•	•	•	•	0	0	2	4	7	14	24	42	68	105	181	297		
130-199	•	•	•	•	•	•	•	•	•	•	•	0	0	2	4	7	13	25	42	72	115	177	301	490		
200-319	•	•	•	•	•	•	•	•	•	•	0	0	2	4	8	14	22	40	68	115	181	277	471			
320-499	•	•	•	•	•	•	•	•	•	0	0	1	4	8	14	24	39	68	113	189						
500-799	•	•	•	•	•	•	•	•	0	0	2	3	7	14	25	40	63	110	181							
800-1249	•	•	•	•	•	•	•	0	0	2	4	7	14	24	42	68	105	181								
1250-1999	•	•	•	•	•	•	0	0	2	4	7	13	24	40	69	110	169									
2000-3149	•	•	•	•	•	0	0	2	4	8	14	22	40	68	115	181										
3150-4999	•	•	•	•	0	0	1	4	8	14	24	38	67	111	186											
5000-7999	•	•	•	0	0	2	3	7	14	25	40	63	110	181												
8000-12499	•	•	0	0	2	4	7	14	24	42	68	105	181													
12500-19999	•	0	0	2	4	7	13	24	40	69	110	169														
20000-31499	0	0	2	4	8	14	22	40	68	115	181															
31500-49999	0	1	4	8	14	24	38	67	111	186																
50000 & Over	2	3	7	14	25	40	63	110	181	301																

• Denotes that the number of sample units from the last ten lots or batches is not sufficient for reduced inspection for this AQL. In this instance more than ten lots or batches may be used for the calculation, provided that the lots or batches used are the most recent ones in sequence, that they have all been on normal inspection, and that none has been rejected while on original inspection.

Source: ANSI/ASQC 1.4–1981, *Sampling Procedures and Tables for Inspection by Attributes*, American Society for Quality Control, Milwaukee, Wis., 1981

Table 9.5 Single-Sampling Plans for Reduced Inspection (Master Table)

Sample size code letter	Sample size	Acceptable Quality Levels (reduced inspection)†																									
		0.010	0.015	0.025	0.040	0.065	0.10	0.15	0.25	0.40	0.65	1.0	1.5	2.5	4.0	6.5	10	15	25	40	65	100	150	250	400	650	1000
		Ac Re	Ac Re	Ac Re	Ac Re	Ac Re	Ac Re	Ac Re	Ac Re	Ac Re	Ac Re	Ac Re	Ac Re	Ac Re	Ac Re	Ac Re	Ac Re	Ac Re	Ac Re	Ac Re	Ac Re	Ac Re	Ac Re	Ac Re	Ac Re	Ac Re	Ac Re
A	2	↓	↓	↓	↓	↓	↓	↓	↓	↓	↓	↓	↓	↓	↓	0 1	↓	↓	1 2	2 3	3 4	5 6	7 8	10 11	14 15	21 22	30 31
B	2	↓	↓	↓	↓	↓	↓	↓	↓	↓	↓	↓	↓	↓	0 1	↑	↓	0 2	1 3	2 4	3 5	5 6	7 8	10 11	14 15	21 22	30 31
C	2	↓	↓	↓	↓	↓	↓	↓	↓	↓	↓	↓	↓	0 1	↑	↓	0 2	1 3	1 4	2 5	3 6	5 8	7 10	10 13	14 17	21 24	↑
D	3	↓	↓	↓	↓	↓	↓	↓	↓	↓	↓	↓	0 1	↑	↓	0 2	1 3	1 4	2 5	3 6	5 8	7 10	10 13	14 17	21 24	↑	↑
E	5	↓	↓	↓	↓	↓	↓	↓	↓	↓	↓	0 1	↑	↓	0 2	1 3	1 4	2 5	3 6	5 8	7 10	10 13	14 17	21 24	↑	↑	↑
F	8	↓	↓	↓	↓	↓	↓	↓	↓	↓	0 1	↑	↓	0 2	1 3	1 4	2 5	3 6	5 8	7 10	10 13	↑	↑	↑	↑	↑	↑
G	13	↓	↓	↓	↓	↓	↓	↓	↓	0 1	↑	↓	0 2	1 3	1 4	2 5	3 6	5 8	7 10	10 13	↑	↑	↑	↑	↑	↑	↑
H	20	↓	↓	↓	↓	↓	↓	↓	0 1	↑	↓	0 2	1 3	1 4	2 5	3 6	5 8	7 10	10 13	↑	↑	↑	↑	↑	↑	↑	↑
J	32	↓	↓	↓	↓	↓	↓	0 1	↑	↓	0 2	1 3	1 4	2 5	3 6	5 8	7 10	10 13	↑	↑	↑	↑	↑	↑	↑	↑	↑
K	50	↓	↓	↓	↓	↓	0 1	↑	↓	0 2	1 3	1 4	2 5	3 6	5 8	7 10	10 13	↑	↑	↑	↑	↑	↑	↑	↑	↑	↑
L	80	↓	↓	↓	↓	0 1	↑	↓	0 2	1 3	1 4	2 5	3 6	5 8	7 10	10 13	↑	↑	↑	↑	↑	↑	↑	↑	↑	↑	↑
M	125	↓	↓	↓	0 1	↑	↓	0 2	1 3	1 4	2 5	3 6	5 8	7 10	10 13	↑	↑	↑	↑	↑	↑	↑	↑	↑	↑	↑	↑
N	200	↓	↓	0 1	↑	↓	0 2	1 3	1 4	2 5	3 6	5 8	7 10	10 13	↑	↑	↑	↑	↑	↑	↑	↑	↑	↑	↑	↑	↑
P	315	↓	0 1	↑	↓	0 2	1 3	1 4	2 5	3 6	5 8	7 10	10 13	↑	↑	↑	↑	↑	↑	↑	↑	↑	↑	↑	↑	↑	↑
Q	500	0 1	↑	↓	0 2	1 3	1 4	2 5	3 6	5 8	7 10	10 13	↑	↑	↑	↑	↑	↑	↑	↑	↑	↑	↑	↑	↑	↑	↑
R	800	↑	↑	0 2	1 3	1 4	2 5	3 6	5 8	7 10	10 13	↑	↑	↑	↑	↑	↑	↑	↑	↑	↑	↑	↑	↑	↑	↑	↑

↓ = Use first sampling plan below arrow. If sample size equals or exceeds lot or batch size, do 100 percent inspection.
↑ = Use first sampling plan above arrow.
Ac = Acceptance number.
Re = Rejection number.
† = If the acceptance number has been exceeded, but the rejection number has not been reached, accept the lot, but reinstate normal inspection (see 10.1.4).

Source: ANSI/ASQC Z1.4–1981, *Sampling Procedures and Tables for Inspection by Attributes*, American Society for Quality Control, Milwaukee, Wis., 1981.

Double and Multiple Sampling

The ANSI Standard contains procedures for double and multiple sampling as well as single sampling. Note that the OC curves for a given lot size and AQL in single, double, and multiple sampling are essentially the same, thereby affording the producer and consumer the same degree of protection. However, once a sampling plan is decided upon, it is not permissible to shift from one to another during evaluation of one lot, since this will result in a different OC curve and therefore an unknown level of protection.

In using the double-sampling tables, Table 9.6 (Table III-A of MIL-Std 105D) for the lot of 1000 and an AQL of 1.0%, the sampling procedure is as follows: Select the first sample of 50 items; accept the lot if zero nonconformances are found and reject the lot if three or more nonconformances are found. Inspect the entire first sample to avoid bias in calculating the process average. If one or two nonconformances are found, inspect the second sample of 50 items. If in the combined first and second sample, three or fewer nonconformances are found, accept the lot. If during the inspection of the second sample, four or more defects are found in the first and second sample combined, inspection may cease and the lot is rejected. It is not necessary to complete the second sample; ANSI Standard Z1.4 states that the process average is calculated based on all samples from original inspection lots; however, bias is introduced when doing so. The previous revision of MIL-Std 105 required that only first-sample results be used in the calculation of the process average, and this seems more reasonable. Criteria similar to that used for single sampling may be used for going to tightened or reduced inspection; refer to Tables 9.7 and 9.8 (Tables III-B and III-C of MIL-Std 105D).

Double sampling will always result in less sampling than single sampling when using MIL-Std 105D. The determination of which to use should be based on this fact as well as other information, such as the cost of getting the sample, likelihood of lot disposition on first sample, need for immediate decision, and other external factors. See the graphs in Table 9.9 (Table IX of MIL-Std 105D) for a comparison of sample sizes among single sampling (the top horizontal line on each graph), double sampling, and multiple sampling.

Multiple sampling is an extension of double sampling except that it provides for seven samples rather than two. It requires fewer samples on the average than double sampling but is more complex to implement and administer. Its use is, therefore, not as widespread.

Table 9.6 Double-Sampling Plans for Normal Inspection (Master Table)

Sample size code letter	Sample	Sample size	Cumulative sample size	Acceptable Quality Levels (normal inspection)																									
				0.010	0.015	0.025	0.040	0.065	0.10	0.15	0.25	0.40	0.65	1.0	1.5	2.5	4.0	6.5	10	15	25	40	65	100	150	250	400	650	1000
				Ac Re	Ac Re	Ac Re	Ac Re	Ac Re	Ac Re	Ac Re	Ac Re	Ac Re	Ac Re	Ac Re	Ac Re	Ac Re	Ac Re	Ac Re	Ac Re	Ac Re	Ac Re	Ac Re	Ac Re	Ac Re	Ac Re	Ac Re	Ac Re	Ac Re	Ac Re
A				↓	↓	↓	↓	↓	↓	↓	↓	↓	↓	↓	↓	↓	↓	•	↓	↓	•	•	•	•	•	•	•	•	•
B	First	2	2	↓	↓	↓	↓	↓	↓	↓	↓	↓	↓	↓	↓	↓	•	↑	↓	0 2	0 3	1 4	2 5	3 7	5 9	7 11	11 16	17 22	25 31
	Second	2	4																	1 2	3 4	4 5	6 7	8 9	12 13	18 19	26 27	37 38	56 57
C	First	3	3	↓	↓	↓	↓	↓	↓	↓	↓	↓	↓	↓	↓	•	↑	↓	0 2	0 3	1 4	2 5	3 7	5 9	7 11	11 16	17 22	25 31	↑
	Second	3	6																1 2	3 4	4 5	6 7	8 9	12 13	18 19	26 27	37 38	56 57	
D	First	5	5	↓	↓	↓	↓	↓	↓	↓	↓	↓	↓	↓	•	↑	↓	0 2	0 3	1 4	2 5	3 7	5 9	7 11	11 16	17 22	25 31	↑	↑
	Second	5	10															1 2	3 4	4 5	6 7	8 9	12 13	18 19	26 27	37 38	56 57		
E	First	8	8	↓	↓	↓	↓	↓	↓	↓	↓	↓	↓	•	↑	↓	0 2	0 3	1 4	2 5	3 7	5 9	7 11	11 16	17 22	25 31	↑	↑	↑
	Second	8	16														1 2	3 4	4 5	6 7	8 9	12 13	18 19	26 27	37 38	56 57			
F	First	13	13	↓	↓	↓	↓	↓	↓	↓	↓	↓	•	↑	↓	0 2	0 3	1 4	2 5	3 7	5 9	7 11	11 16	↑	↑	↑	↑	↑	↑
	Second	13	26													1 2	3 4	4 5	6 7	8 9	12 13	18 19	26 27						
G	First	20	20	↓	↓	↓	↓	↓	↓	↓	↓	•	↑	↓	0 2	0 3	1 4	2 5	3 7	5 9	7 11	11 16	↑	↑	↑	↑	↑	↑	↑
	Second	20	40												1 2	3 4	4 5	6 7	8 9	12 13	18 19	26 27							
H	First	32	32	↓	↓	↓	↓	↓	↓	↓	•	↑	↓	0 2	0 3	1 4	2 5	3 7	5 9	7 11	11 16	↑	↑	↑	↑	↑	↑	↑	↑
	Second	32	64											1 2	3 4	4 5	6 7	8 9	12 13	18 19	26 27								
J	First	50	50	↓	↓	↓	↓	↓	↓	•	↑	↓	0 2	0 3	1 4	2 5	3 7	5 9	7 11	11 16	↑	↑	↑	↑	↑	↑	↑	↑	↑
	Second	50	100										1 2	3 4	4 5	6 7	8 9	12 13	18 19	26 27									
K	First	80	80	↓	↓	↓	↓	↓	•	↑	↓	0 2	0 3	1 4	2 5	3 7	5 9	7 11	11 16	↑	↑	↑	↑	↑	↑	↑	↑	↑	↑
	Second	80	160									1 2	3 4	4 5	6 7	8 9	12 13	18 19	26 27										
L	First	125	125	↓	↓	↓	↓	•	↑	↓	0 2	0 3	1 4	2 5	3 7	5 9	7 11	11 16	↑	↑	↑	↑	↑	↑	↑	↑	↑	↑	↑
	Second	125	250								1 2	3 4	4 5	6 7	8 9	12 13	18 19	26 27											
M	First	200	200	↓	↓	↓	•	↑	↓	0 2	0 3	1 4	2 5	3 7	5 9	7 11	11 16	↑	↑	↑	↑	↑	↑	↑	↑	↑	↑	↑	↑
	Second	200	400							1 2	3 4	4 5	6 7	8 9	12 13	18 19	26 27												
N	First	315	315	↓	↓	•	↑	↓	0 2	0 3	1 4	2 5	3 7	5 9	7 11	11 16	↑	↑	↑	↑	↑	↑	↑	↑	↑	↑	↑	↑	↑
	Second	315	630						1 2	3 4	4 5	6 7	8 9	12 13	18 19	26 27													
P	First	500	500	↓	•	↑	↓	0 2	0 3	1 4	2 5	3 7	5 9	7 11	11 16	↑	↑	↑	↑	↑	↑	↑	↑	↑	↑	↑	↑	↑	↑
	Second	500	1000					1 2	3 4	4 5	6 7	8 9	12 13	18 19	26 27														
Q	First	800	800	•	↑	↓	0 2	0 3	1 4	2 5	3 7	5 9	7 11	11 16	↑	↑	↑	↑	↑	↑	↑	↑	↑	↑	↑	↑	↑	↑	↑
	Second	800	1600				1 2	3 4	4 5	6 7	8 9	12 13	18 19	26 27															
R	First	1250	1250	↑	↑	0 2	0 3	1 4	2 5	3 7	5 9	7 11	11 16	↑	↑	↑	↑	↑	↑	↑	↑	↑	↑	↑	↑	↑	↑	↑	↑
	Second	1250	2500			1 2	3 4	4 5	6 7	8 9	12 13	18 19	26 27																

↓ = Use first sampling plan below arrow. If sample size equals or exceeds lot or batch size, do 100 percent inspection.
↑ = Use first sampling plan above arrow.
Ac = Acceptance number
Re = Rejection number
• = Use corresponding single sampling plan (or alternatively, use double sampling plan below, where available).

Source: ANSI/ASQC Z1.4–1981, *Sampling Procedures and Tables for Inspection by Attributes*, American Society for Quality Control, Milwaukee, Wis., 1981.

Table 9.7 Double-Sampling Plans for Tightened Inspection (Master Table)

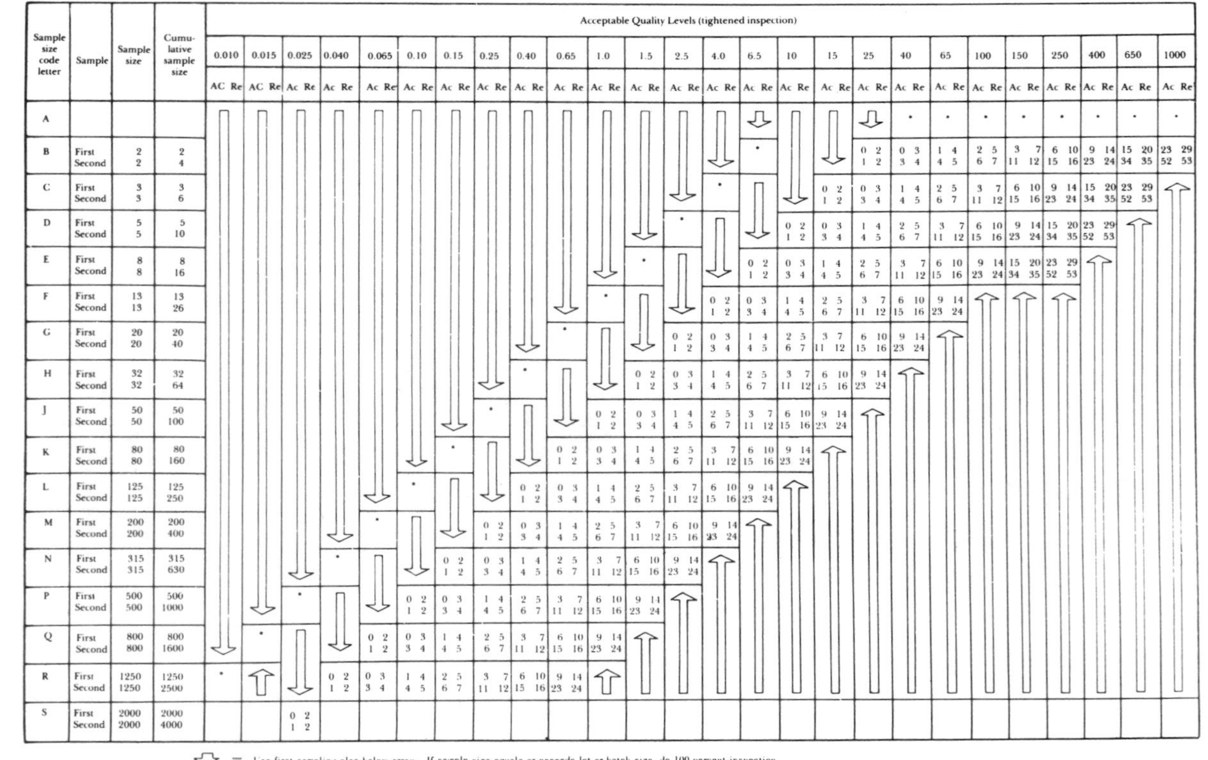

Acceptable Quality Levels (tightened inspection)

Sample size code letter	Sample	Sample size	Cumulative sample size	0.010	0.015	0.025	0.040	0.065	0.10	0.15	0.25	0.40	0.65
				AC Re	AC Re	Ac Re	Ac Re	Ac Re	Ac Re	Ac Re	Ac Re	Ac Re	Ac Re
A				↓	↓	↓	↓	↓	↓	↓	↓	↓	↓
B	First Second	2 2	2 4	↓	↓	↓	↓	↓	↓	↓	↓	↓	↓
C	First Second	3 3	3 6	↓	↓	↓	↓	↓	↓	↓	↓	↓	↓
D	First Second	5 5	5 10	↓	↓	↓	↓	↓	↓	↓	↓	↓	↓
E	First Second	8 8	8 16	↓	↓	↓	↓	↓	↓	↓	↓	↓	↓
F	First Second	13 13	13 26	↓	↓	↓	↓	↓	↓	↓	↓	↓	↓
G	First Second	20 20	20 40	↓	↓	↓	↓	↓	↓	↓	↓	↓	•
H	First Second	32 32	32 64	↓	↓	↓	↓	↓	↓	↓	↓	•	↓
J	First Second	50 50	50 100	↓	↓	↓	↓	↓	↓	↓	•	↓	↓
K	First Second	80 80	80 160	↓	↓	↓	↓	↓	↓	•	↓	↓	0 2 1 2
L	First Second	125 125	125 250	↓	↓	↓	↓	↓	•	↓	↓	0 2 1 2	0 3 3 4
M	First Second	200 200	200 400	↓	↓	↓	↓	•	↓	↓	0 2 1 2	0 3 3 4	1 4 4 5
N	First Second	315 315	315 630	↓	↓	↓	•	↓	↓	0 2 1 2	0 3 3 4	1 4 4 5	2 5 6 7
P	First Second	500 500	500 1000	↓	↓	•	↓	↓	0 2 1 2	0 3 3 4	1 4 4 5	2 5 6 7	3 7 11 12
Q	First Second	800 800	800 1600	↓	•	↓	↓	0 2 1 2	0 3 3 4	1 4 4 5	2 5 6 7	3 7 11 12	6 10 15 16
R	First Second	1250 1250	1250 2500	•	↑	↓	0 2 1 2	0 3 3 4	1 4 4 5	2 5 6 7	3 7 11 12	6 10 15 16	9 14 23 24
S	First Second	2000 2000	2000 4000			0 2 1 2							

Sample size code letter	Sample	Sample size	Cumulative sample size	1.0	1.5	2.5	4.0	6.5	10	15	25	40	65	100	150	250	400	650	1000
				Ac Re	Ac Re	Ac Re	Ac Re	Ac Re	Ac Re	Ac Re	Ac Re	Ac Re	Ac Re	Ac Re	Ac Re	Ac Re	Ac Re	Ac Re	Ac Re
A				↓	↓	↓	↓	↓	↓	↓	↓	•	•	•	•	•	•	•	•
B	First Second	2 2	2 4	↓	↓	↓	↓	•	↓	↓	0 2 1 2	0 3 3 4	1 4 4 5	2 5 6 7	3 7 11 12	6 10 15 16	9 14 23 24	15 20 34 35	23 29 52 53
C	First Second	3 3	3 6	↓	↓	↓	•	↓	↓	0 2 1 2	0 3 3 4	1 4 4 5	2 5 6 7	3 7 11 12	6 10 15 16	9 14 23 24	15 20 34 35	23 29 52 53	↑
D	First Second	5 5	5 10	↓	↓	•	↓	↓	0 2 1 2	0 3 3 4	1 4 4 5	2 5 6 7	3 7 11 12	6 10 15 16	9 14 23 24	15 20 34 35	23 29 52 53	↑	↑
E	First Second	8 8	8 16	↓	•	↓	↓	0 2 1 2	0 3 3 4	1 4 4 5	2 5 6 7	3 7 11 12	6 10 15 16	9 14 23 24	15 20 34 35	23 29 52 53	↑	↑	↑
F	First Second	13 13	13 26	•	↓	↓	0 2 1 2	0 3 3 4	1 4 4 5	2 5 6 7	3 7 11 12	6 10 15 16	9 14 23 24	↑	↑	↑	↑	↑	↑
G	First Second	20 20	20 40	↓	↓	0 2 1 2	0 3 3 4	1 4 4 5	2 5 6 7	3 7 11 12	6 10 15 16	9 14 23 24	↑	↑	↑	↑	↑	↑	↑
H	First Second	32 32	32 64	↓	0 2 1 2	0 3 3 4	1 4 4 5	2 5 6 7	3 7 11 12	6 10 15 16	9 14 23 24	↑	↑	↑	↑	↑	↑	↑	↑
J	First Second	50 50	50 100	0 2 1 2	0 3 3 4	1 4 4 5	2 5 6 7	3 7 11 12	6 10 15 16	9 14 23 24	↑	↑	↑	↑	↑	↑	↑	↑	↑
K	First Second	80 80	80 160	0 3 3 4	1 4 4 5	2 5 6 7	3 7 11 12	6 10 15 16	9 14 23 24	↑	↑	↑	↑	↑	↑	↑	↑	↑	↑
L	First Second	125 125	125 250	1 4 4 5	2 5 6 7	3 7 11 12	6 10 15 16	9 14 23 24	↑	↑	↑	↑	↑	↑	↑	↑	↑	↑	↑
M	First Second	200 200	200 400	2 5 6 7	3 7 11 12	6 10 15 16	9 14 23 24	↑	↑	↑	↑	↑	↑	↑	↑	↑	↑	↑	↑
N	First Second	315 315	315 630	3 7 11 12	6 10 15 16	9 14 23 24	↑	↑	↑	↑	↑	↑	↑	↑	↑	↑	↑	↑	↑
P	First Second	500 500	500 1000	6 10 15 16	9 14 23 24	↑	↑	↑	↑	↑	↑	↑	↑	↑	↑	↑	↑	↑	↑
Q	First Second	800 800	800 1600	9 14 23 24	↑	↑	↑	↑	↑	↑	↑	↑	↑	↑	↑	↑	↑	↑	↑
R	First Second	1250 1250	1250 2500	↑	↑	↑	↑	↑	↑	↑	↑	↑	↑	↑	↑	↑	↑	↑	↑
S	First Second	2000 2000	2000 4000																

↓ = Use first sampling plan below arrow. If sample size equals or exceeds lot or batch size, do 100 percent inspection.
↑ = Use first sampling plan above arrow
Ac = Acceptance number
Re = Rejection number
• = Use corresponding single sampling plan (or, alternatively, use double sampling plan below, where available).

Source: ANSI/ASQC Z1.4–1981, *Sampling Procedures and Tables for Inspection by Attributes*, American Society for Quality Control, Milwaukee, Wis., 1981.

Table 9.8 Double-Sampling Plans for Reduced Inspection (Master Table)

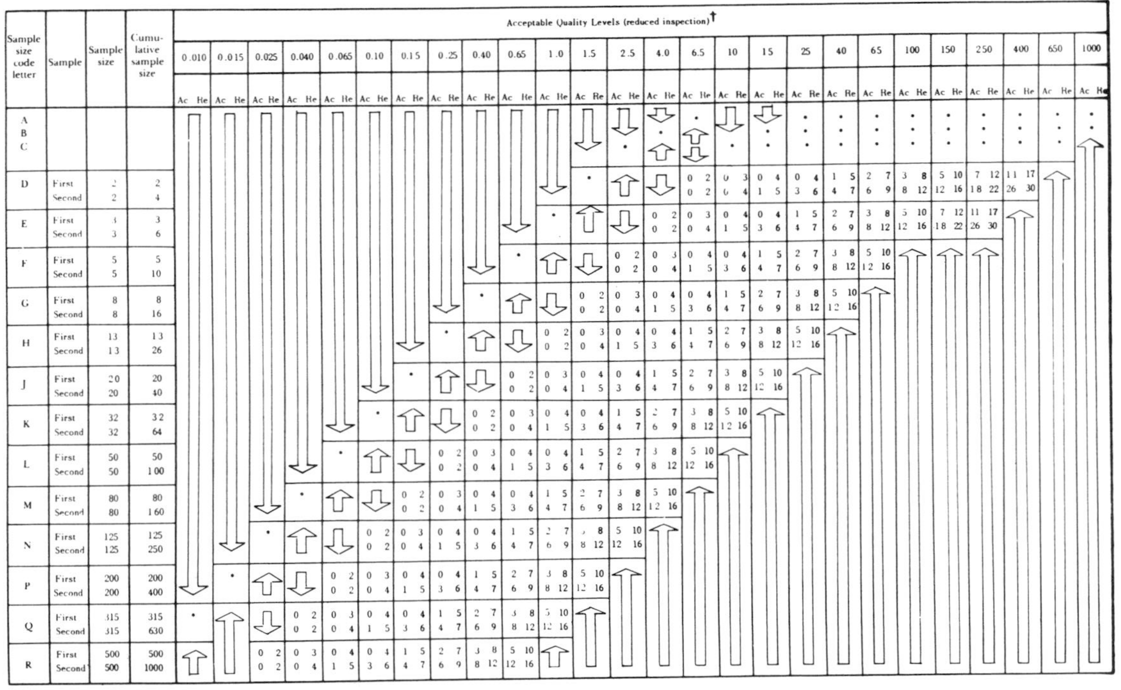

Sample size code letter	Sample	Sample size	Cumulative sample size	Acceptable Quality Levels (reduced inspection)†																									
				0.010	0.015	0.025	0.040	0.065	0.10	0.15	0.25	0.40	0.65	1.0	1.5	2.5	4.0	6.5	10	15	25	40	65	100	150	250	400	650	1000
				Ac Re	Ac Re	Ac Re	Ac Re	Ac Re	Ac Re	Ac Re	Ac Re	Ac Re	Ac Re	Ac Re	Ac Re	Ac Re	Ac Re	Ac Re	Ac Re	Ac Re	Ac Re	Ac Re	Ac Re	Ac Re	Ac Re	Ac Re	Ac Re	Ac Re	Ac Re
A				↓	↓	↓	↓	↓	↓	↓	↓	↓	↓	↓	↓	↓	↓	•	↓	↓	•	•	•	•	•	•	•	•	•
B				↓	↓	↓	↓	↓	↓	↓	↓	↓	↓	↓	↓	↓	•	↑	↓	•	•	•	•	•	•	•	•	•	•
C				↓	↓	↓	↓	↓	↓	↓	↓	↓	↓	↓	↓	•	↑	↓	•	•	•	•	•	•	•	•	•	•	↑
D	First	2	2	↓	↓	↓	↓	↓	↓	↓	↓	↓	↓	↓	•	↑	↓	0 2	0 3	0 4	0 4	1 5	2 7	3 8	5 10	7 12	11 17	↑	↑
	Second	2	4															0 2	0 4	1 5	3 6	4 7	6 9	8 12	12 16	18 22	26 30		
E	First	3	3	↓	↓	↓	↓	↓	↓	↓	↓	↓	↓	•	↑	↓	0 2	0 3	0 4	0 4	1 5	2 7	3 8	5 10	7 12	11 17	↑	↑	↑
	Second	3	6														0 2	0 4	1 5	3 6	4 7	6 9	8 12	12 16	18 22	26 30			
F	First	5	5	↓	↓	↓	↓	↓	↓	↓	↓	↓	•	↑	↓	0 2	0 3	0 4	0 4	1 5	2 7	3 8	5 10	↑	↑	↑	↑	↑	↑
	Second	5	10													0 2	0 4	1 5	3 6	4 7	6 9	8 12	12 16						
G	First	8	8	↓	↓	↓	↓	↓	↓	↓	↓	•	↑	↓	0 2	0 3	0 4	0 4	1 5	2 7	3 8	5 10	↑	↑	↑	↑	↑	↑	↑
	Second	8	16											0 2	0 4	1 5	3 6	4 7	6 9	8 12	12 16								
H	First	13	13	↓	↓	↓	↓	↓	↓	↓	•	↑	↓	0 2	0 3	0 4	0 4	1 5	2 7	3 8	5 10	↑	↑	↑	↑	↑	↑	↑	↑
	Second	13	26											0 2	0 4	1 5	3 6	4 7	6 9	8 12	12 16								
J	First	20	20	↓	↓	↓	↓	↓	↓	•	↑	↓	0 2	0 3	0 4	0 4	1 5	2 7	3 8	5 10	↑	↑	↑	↑	↑	↑	↑	↑	↑
	Second	20	40										0 2	0 4	1 5	3 6	4 7	6 9	8 12	12 16									
K	First	32	32	↓	↓	↓	↓	↓	•	↑	↓	0 2	0 3	0 4	0 4	1 5	2 7	3 8	5 10	↑	↑	↑	↑	↑	↑	↑	↑	↑	↑
	Second	32	64									0 2	0 4	1 5	3 6	4 7	6 9	8 12	12 16										
L	First	50	50	↓	↓	↓	↓	•	↑	↓	0 2	0 3	0 4	0 4	1 5	2 7	3 8	5 10	↑	↑	↑	↑	↑	↑	↑	↑	↑	↑	↑
	Second	50	100								0 2	0 4	1 5	3 6	4 7	6 9	8 12	12 16											
M	First	80	80	↓	↓	↓	•	↑	↓	0 2	0 3	0 4	0 4	1 5	2 7	3 8	5 10	↑	↑	↑	↑	↑	↑	↑	↑	↑	↑	↑	↑
	Second	80	160							0 2	0 4	1 5	3 6	4 7	6 9	8 12	12 16												
N	First	125	125	↓	↓	•	↑	↓	0 2	0 3	0 4	0 4	1 5	2 7	3 8	5 10	↑	↑	↑	↑	↑	↑	↑	↑	↑	↑	↑	↑	↑
	Second	125	250						0 2	0 4	1 5	3 6	4 7	6 9	8 12	12 16													
P	First	200	200	↓	•	↑	↓	0 2	0 3	0 4	0 4	1 5	2 7	3 8	5 10	↑	↑	↑	↑	↑	↑	↑	↑	↑	↑	↑	↑	↑	↑
	Second	200	400					0 2	0 4	1 5	3 6	4 7	6 9	8 12	12 16														
Q	First	315	315	•	↑	↓	0 2	0 3	0 4	0 4	1 5	2 7	3 8	5 10	↑	↑	↑	↑	↑	↑	↑	↑	↑	↑	↑	↑	↑	↑	↑
	Second	315	630				0 2	0 4	1 5	3 6	4 7	6 9	8 12	12 16															
R	First	500	500	↑	↑	0 2	0 3	0 4	0 4	1 5	2 7	3 8	5 10	↑	↑	↑	↑	↑	↑	↑	↑	↑	↑	↑	↑	↑	↑	↑	↑
	Second	500	1000			0 2	0 4	1 5	3 6	4 7	6 9	8 12	12 16																

↓ = Use first sampling plan below arrow. If sample size equals or exceeds lot or batch size, do 100 percent inspection.
↑ = Use first sampling plan above arrow.
Ac = Acceptance number.
Re = Rejection number.
• = Use corresponding single sampling plan (or alternatively, use double sampling plan below, when available.)
† = If, after the second sample, the acceptance number has been exceeded, but the rejection number has not been reached, accept the lot, but reinstate normal inspection (see 10.14).

Source: ANSI/ASQC Z1.4–1981, *Sampling Procedures and Tables for Inspection by Attributes*, American Society for Quality Control, Milwaukee, Wis., 1981.

Table 9.9 Average Sample Size Curves for Double and Multiple Sampling

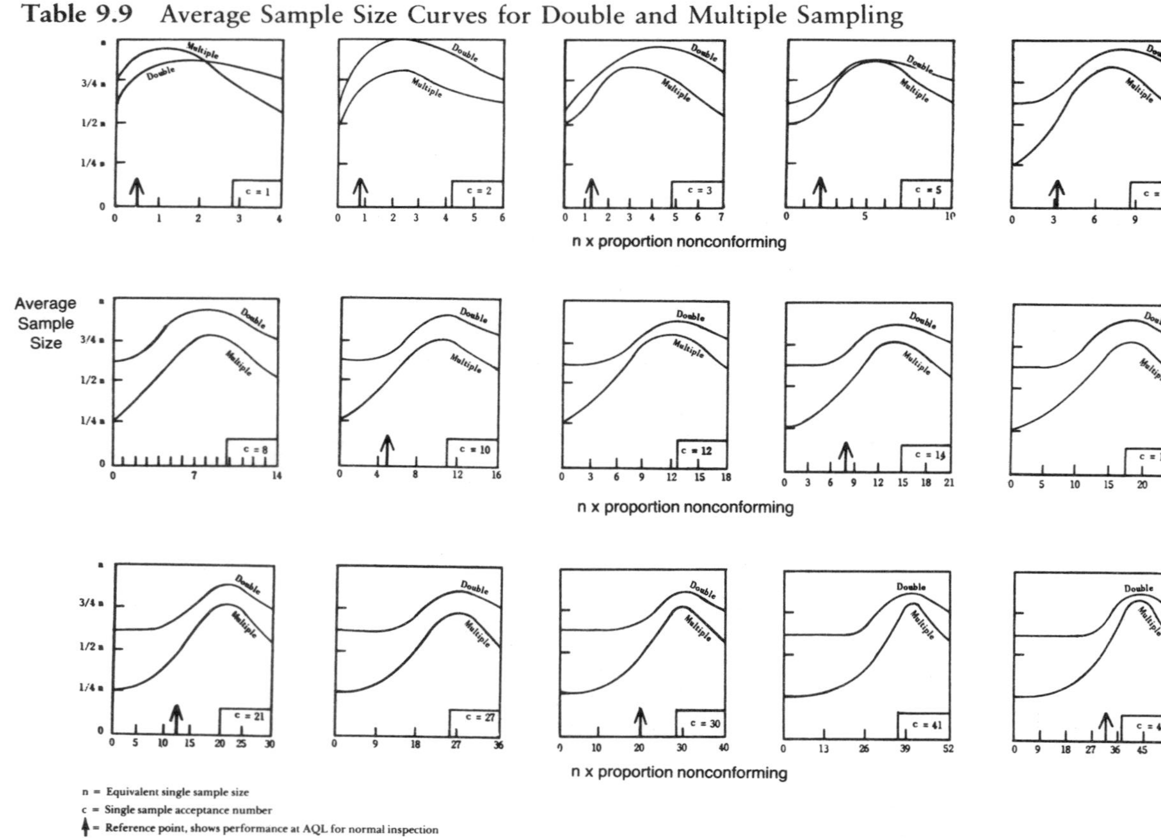

Source: ANSI/ASQC Z1.4–1981, *Sampling Procedures and Tables for Inspection by Attributes*, American Society for Quality Control, Milwaukee, Wis., 1981.

When a lot is rejected, it must be rescreened for the failing category of nonconformities or the particular item of rejection before resubmission for reinspection, depending upon preestablished rules. Frequently, however, the question of rescreening for all categories is raised because this may be desirable from a practical viewpoint. These decisions can be made on a case-by-case basis.

In ANSI Standard Z1.4 the OC curves are tabulated by AQL. Tables for each code letter are provided in the standard showing sampling plans for single, double, and multiple sampling, along with OC curves and tabulated values for each AQL value. They are not included here. A complete set of plans available for one code letter, Tables 9.10 and 9.11 (Table X-J of MIL-Std 105D), is included as a sample.

Dodge–Romig Tables

In the Dodge–Romig Tables, single- and double-sampling plans are classified by AOQL from 0.1% to 10% and by LTPD in the range 0.5 to 10%. Considering the AOQL tables first, refer to Table 9.12; the plan will vary with the process average. In the absence of any information concerning the process average, the poorest level must be considered, so the sampling plan will be $n = 120$, $c = 2$, where n is the sample size and c is the allowable number of defects. As information is collected enabling the calculation of the estimated process average, sampling may be reduced to a sampling plan with $n = 35$, $c = 0$, a significant reduction in the cost of sampling, depending on the value calculated as the process average.

The value for the AOQL is based on removal of all the defects in a rejected lot through 100% reinspection and replacement with good parts and inclusion of this lot with the remainder of the product being shipped; therefore, its contribution to the average outgoing quality level will be toward improving or lowering the percent defective. In those instances where lots that fail sampling are diverted and not rescreened, the AOQL concept will not truly reflect the limiting value of the percent defective contained in shipped lots.

In using LTPD tables, the consideration is placed on the protection given for each individual lot. The LTPD value is that percent defective at which the probability of acceptance is 10%. This sampling is more common if the individual lot quality is critical, rather than the AOQL, which provides average protection. Using the same lot size of 1000 with a 1% LTPD plan, Table 9.13; the sampling plan for the poorest process average will be $n = 335$, $c = 1$ (considerably greater than $n = 120$,

Table 9.10 Tables for Sample Size Code Letter: J

INDIVIDUAL PLANS

CHART J - OPERATING CHARACTERISTIC CURVES FOR SINGLE SAMPLING PLANS

(Curves for double and multiple sampling are matched as closely as practicable)

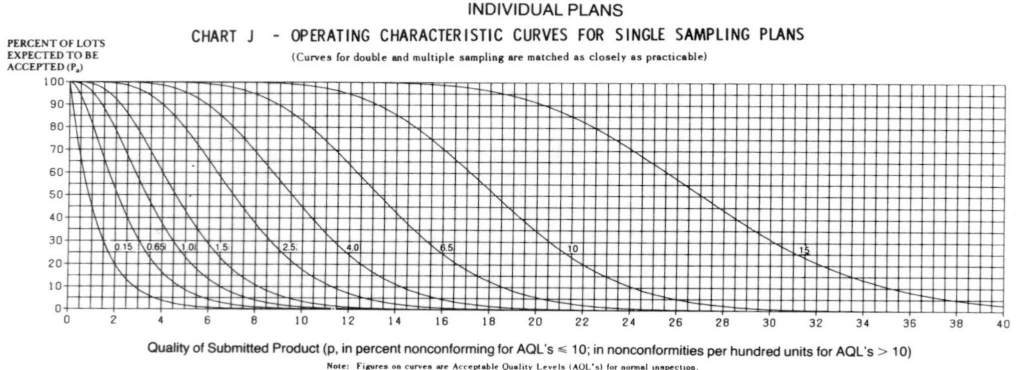

Note: Figures on curves are Acceptable Quality Levels (AQL's) for normal inspection.

TABLE X-J-1 - TABULATED VALUES FOR OPERATING CHARACTERISTIC CURVES FOR SINGLE SAMPLING PLANS

P_a	Acceptable Quality Levels (normal inspection)																					
	0.15	0.65	1.0	1.5	2.5	4.0	X	6.5	X	10	0.15	0.65	1.0	1.5	2.5	4.0	X	6.5	X	10	X	15
	p (in percent nonconforming)										p (in nonconformities per hundred units)											
99.0	0.013	0.188	0.550	1.05	2.30	3.72	4.50	6.13	7.88	9.75	0.013	0.186	0.545	1.03	2.23	3.63	4.38	5.96	7.62	9.35	12.9	15.7
95.0	0.064	0.444	1.03	1.73	3.32	5.06	5.98	7.91	9.89	11.9	0.064	0.444	1.02	1.71	3.27	4.98	5.87	7.71	9.61	11.6	15.6	18.6
90.0	0.132	0.666	1.38	2.20	3.98	5.91	6.91	8.95	11.0	13.2	0.131	0.665	1.38	2.18	3.94	5.82	6.79	8.78	10.8	12.9	17.1	20.3
75.0	0.359	1.202	2.16	3.18	5.30	7.50	8.62	10.9	13.2	15.5	0.360	1.20	2.16	3.17	5.27	7.45	8.55	10.8	13.0	15.3	19.9	23.4
50.0	0.863	2.09	3.33	4.57	7.06	9.55	10.8	13.3	15.8	18.3	0.866	2.10	3.34	4.59	7.09	9.59	10.8	13.3	15.8	18.3	23.3	27.1
25.0	1.72	3.33	4.84	6.31	9.14	11.9	13.3	16.0	18.6	21.3	1.73	3.37	4.90	6.39	9.28	12.1	13.5	16.3	19.0	21.8	27.2	31.2
10.0	2.84	4.78	6.52	8.16	11.3	14.2	15.7	18.6	21.4	24.2	2.88	4.86	6.65	8.35	11.6	14.7	16.2	19.3	22.2	25.2	30.9	35.2
5.0	3.68	5.80	7.66	9.39	12.7	15.8	17.3	20.3	23.2	26.0	3.75	5.93	7.87	9.69	13.1	16.4	18.0	21.2	24.3	27.4	33.4	37.8
1.0	5.59	8.00	10.1	12.0	15.6	18.9	20.5	23.6	26.5	29.5	5.76	8.30	10.5	12.6	16.4	20.0	21.8	25.2	28.5	31.8	38.2	42.9
	0.25	1.0	1.5	2.5	4.0	X	6.5	X	10	X	0.25	1.0	1.5	2.5	4.0	X	6.5	X	10	X	15	X
	Acceptable Quality Levels (tightened inspection)																					

Note: Binomial distribution used for percent nonconforming computations; Poisson for nonconformities per hundred units.

Source: ANSI/ASQC Z1.4–1981, *Sampling Procedures and Tables for Inspection by Attributes*, American Society for Quality Control, Milwaukee, Wis., 1981.

Table 9.11 Sampling Plans for Sample Size Code Letter: J

Acceptable Quality Levels (normal inspection)

Type of sampling plan	Cumulative sample size	Less than 0.15		0.15		0.25		X		0.40		0.65		1.0		1.5		2.5		4.0		X		6.5		X		10		X		15		Higher than 15		Cumulative sample size
		Ac	Re	Ac	Re	Ac	Rc	Ac	Re	Ac	Re	Ac	Re	Ac	Re	Ac	Re	Ac	Re	Ac	Re	Ac	Re	Ac	Re	Ac	Re	Ac	Re	Ac	Re	Ac	Re	Ac	Re	
Single	80	∇		0	1	Use Letter H		Use Letter L		Use Letter K		1	2	2	3	3	4	5	6	7	8	8	9	10	11	12	13	14	15	18	19	21	22	△		80
Double	50	∇		•								0	2	0	3	1	4	2	5	3	7	3	7	5	9	6	10	7	11	9	14	11	16	△		50
	100											1	2	3	4	4	5	6	7	8	9	11	12	12	13	15	16	18	19	23	24	26	27			100
Multiple	20	∇		•								#	2	#	2	#	3	#	4	0	4	0	4	0	5	0	6	1	7	1	8	2	9	△		20
	40											#	2	0	3	0	3	1	5	1	6	2	7	3	8	3	9	4	10	6	12	7	14			40
	60											0	2	0	3	1	4	2	6	3	8	4	9	6	10	7	12	8	13	11	17	13	19			60
	80											0	3	1	4	2	5	3	7	5	10	6	11	8	13	10	15	12	17	16	22	19	25			80
	100											1	3	2	4	3	6	5	8	7	11	9	12	11	15	14	17	17	20	22	25	25	29			100
	120											1	3	3	5	4	6	7	9	10	12	12	14	14	17	18	20	21	23	27	29	31	33			120
	140											2	3	4	5	6	7	9	10	13	14	14	15	18	19	21	22	25	26	32	33	37	38			140
		Less than 0.25		0.25		X		0.40		0.65		1.0		1.5		2.5		4.0		X		6.5		X		10		X		15		X		Higher than 15		

Acceptable Quality Levels (tightened inspection)

△ = Use next preceding sample size code letter for which acceptance and rejection numbers are available.
∇ = Use next subsequent sample size code letter for which acceptance and rejection numbers are available.
Ac = Acceptance number
Re = Rejection number
• = Use single sampling plan above (or alternatively use letter M)
= Acceptance not permitted at this sample size.

Source: ANSI/ASQC Z1.4–1981, *Sampling Procedures and Tables for Inspection by Attributes*, American Society for Quality Control, Milwaukee, Wis., 1981.

Table 9.12 Single-Sampling Table for Average Outgoing Quality Limit (AOQL) = 1.0%

	Process Average																	
	0–0.02%			*0.03–0.20%*			*0.21–0.40%*			*0.41–0.60%*			*0.61–0.80%*			*0.81–1.00%*		
Lot size	n	c	p_t %	n	c	p_t %	n	c	p_t %	n	c	p_t %	n	c	p_t %	n	c	p_t %
1– 25	All	0		All	0		All	0		All	0		All	0		All	0	
26– 50	22	0	7.7	22	0	7.7	22	0	7.7	22	0	7.7	22	0	7.7	22	0	7.7
51– 100	27	0	7.1	27	0	7.1	27	0	7.1	27	0	7.1	27	0	7.1	27	0	7.1
101– 200	32	0	6.4	32	0	6.4	32	0	6.4	32	0	6.4	32	0	6.4	32	0	6.4
201– 300	33	0	6.3	33	0	6.3	33	0	6.3	33	0	6.3	33	0	6.3	65	1	5.0
301– 400	34	0	6.1	34	0	6.1	34	0	6.1	70	1	4.6	70	1	4.6	70	1	4.6
401– 500	35	0	6.1	35	0	6.1	75	1	4.4	75	1	4.4	75	1	4.4	75	1	4.4
601– 800	35	0	6.2	35	0	6.2	75	1	4.4	75	1	4.4	75	1	4.4	120	2	4.2
801– 1,000	35	0	6.3	34	0	6.3	80	1	4.4	80	1	4.4	120	2	4.3	120	2	4.3
1,001– 2,000	36	0	6.2	80	1	4.5	80	1	4.5	130	2	4.0	130	2	4.0	180	3	3.7
2,001– 3,000	36	0	6.2	80	1	4.6	80	1	4.6	130	2	4.0	185	3	3.6	235	4	3.3
3,001– 4,000	36	0	6.2	80	1	4.7	135	2	3.9	135	2	3.9	185	3	3.6	295	5	3.1
4,001– 5,000	36	0	6.2	85	1	4.6	135	2	3.9	190	3	3.5	245	4	3.2	300	5	3.1
5,001– 7,000	37	0	6.1	85	1	4.6	135	2	3.9	190	3	3.5	305	5	3.0	420	7	2.8
7,001– 10,000	37	0	6.2	85	1	4.6	135	2	3.9	245	4	3.2	310	5	3.0	430	7	2.7
10,001– 20,000	85	1	4.6	135	2	3.9	195	3	3.4	250	4	3.2	435	7	2.7	635	10	2.4
20,001– 50,000	85	1	4.6	135	2	3.9	255	4	3.1	380	6	2.8	575	9	2.5	990	15	2.1
50,001–100,000	85	1	4.6	135	2	3.9	255	4	3.1	445	7	2.6	790	12	2.3	1520	22	1.9

Source: Dodge, H. F. and H. G. Romig (1959). *Sampling Inspection Tables: Single and Double Sampling*, John Wiley and Sons, New York.
[a]*n*, Sample size; c, acceptance number; "All" indicates that each piece in the lot is to be inspected.

Table 9.13 Single-Sampling Table for Lot Tolerance Percent Defective (LTPD) = 1%

Lot size	Process average[a]																	
	0–0.010%			0.011–0.10%			0.11–0.20%			0.21–0.30%			0.31–0.40%			0.41–0.50%		
	n	*c*	AOQL (%)	*n*	*c*	AOQL (%)	*n*	*c*	AOQL (%)	*n*	*c*	AOQL (%)	*n*	*c*	AOQL (%)	*n*	*c*	AOQL (%)
1– 120	All	0	0	All	0	0	All	0	0	All	0	0	All	0	0	All	0	0
121– 150	120	0	0.06	120	0	0.06	120	0	0.06	120	0	0.06	120	0	0.06	120	0	0.06
151– 200	140	0	0.08	140	0	0.08	140	0	0.08	140	0	0.08	140	0	0.08	140	0	0.08
201– 300	165	0	9.10	165	0	0.10	165	0	0.10	165	0	0.10	165	0	0.10	165	0	0.10
301– 400	175	0	0.12	175	0	0.12	175	0	0.12	175	0	0.12	175	0	0.12	175	0	0.12
501– 600	190	0	0.13	190	0	0.13	190	0	0.13	190	0	0.13	190	0	0.13	305	1	0.14
601– 800	200	0	0.14	200	0	0.14	200	0	0.14	330	1	0.15	330	1	0.15	330	1	0.15
801– 1,000	205	0	0.14	205	0	0.14	205	0	0.14	335	1	0.17	335	1	0.17	335	1	0.17
1,001– 2,000	220	0	0.15	220	0	0.15	360	1	0.19	490	2	0.21	490	2	0.21	610	3	0.22
2,001– 3,000	220	0	0.15	375	1	0.20	505	2	0.23	630	3	0.24	745	4	0.26	870	5	0.26
3,001– 4,000	225	0	0.15	380	1	0.20	510	2	0.24	645	3	0.25	880	5	0.28	1000	6	0.29
4,001– 5,000	225	0	0.16	380	1	0.20	520	2	0.24	770	4	0.28	895	5	0.29	1120	7	0.31
5,001– 7,000	230	0	0.15	385	1	0.21	655	3	0.27	780	4	0.29	1020	6	0.32	1260	8	0.34
7,001– 10,000	230	0	0.16	520	2	0.25	660	3	0.28	910	5	0.32	1150	7	0.34	1500	10	0.37
10,001– 20,000	390	1	0.21	525	2	0.26	785	4	0.31	1040	6	0.35	1400	9	0.39	1980	14	0.43
50,001–100,000	390	1	0.21	670	3	0.29	1040	6	0.36	1420	9	0.41	2120	15	0.47	3150	23	0.50

Source: Dodge, H. F. and H. G. Romig (1959). *Sampling Inspection Tables: Single and Double Sampling*, John Wiley and Sons, New York.

[a] n, sample size; c, acceptance number; AOQL, Average Outgoing Quality Limit. "All" indicates that each piece in the lot is to be inspected.

$c = 2$ for the 1% AOQL plan). As the process average decreases, the lowest sample size is $n = 205$ with an acceptance number of 0. For an interesting comparison of OC curves for AQL, AOQL, and LTPD where each is 1%, see Fig. 9.5.

Variables Sampling (ANSI Z1.9 and MIL-Std 414)

Because the concept of variables sampling is important in acceptance sampling, a brief discussion will be included here to indicate areas of possible use. The variables sampling concept is based on selecting a sample from a lot that has a known distribution. Using the values obtained from the sample, an estimate of the average and standard deviation (or range) is obtained. When these values are used to estimate the lot limits and then compared with actual specification values, the percentage of nonconforming items can be estimated. Variables sampling plans are prepared in sampling tables in ANSI Standard Z1.9 (MIL-Std 414) and in other sources, such as Bowker and Goode.

Table 9.14 AQL Conversion Table for Variables Sampling

For specified AWL values falling within these ranges:	Use this AQL value:
0 – 0.049	0.04
0.050 – 0.069	0.065
0.070 – 0.109	0.10
0.110 – 0.164	0.15
0.165 – 0.279	0.25
0.280 – 0.439	0.40
0.440 – 0.699	0.65
0.700 – 1.09	1.0
1.10 – 1.64	1.5
1.65 – 2.79	2.5
2.80 – 4.39	4.0
4.40 – 6.99	6.5
7.00 – 10.9	10.0
11.00 – 16.4	15.0

Source: MIL–Std 414 (1957).

Table 9.15 Sample Size Code Letters for Variables Sampling[a]

Lot Size	*Inspection levels*				
	I	*II*	*III*	*IV*	*V*
3– 8	B	B	B	B	C
9– 15	B	B	B	B	D
16– 25	B	B	B	C	E
26– 40	B	B	B	D	F
41– 65	B	B	C	E	G
66– 110	B	B	D	F	H
111– 180	B	C	E	G	I
181– 300	B	D	F	H	J
301– 500	C	E	G	I	K
501– 800	D	F	H	J	L
801– 1,300	E	G	I	K	L
1,301– 3,200	F	H	J	L	M
3,200– 8,000	G	I	L	M	N
8,001– 22,000	H	J	M	N	O
22,001–110,000	I	K	N	O	P
110,001–550,000	I	K	O	P	Q
550,000 and over	I	K	P	Q	Q

Source: MIL–Std 414 (1957).
[a]Sample size code letters given are applicable when the indicated inspection levels are to be used.

In using ANSI Standard Z1.9 for a 1% AQL value for lot sizes of approximately 1000, the sample size is 35. This is determined by using Table 9.14 (Table A-1 in MIL-Std 414) and Table 9.15 (Table A-2 in MIL-Std 414). Inspection level IV is always used unless otherwise specified. Then Table 9.16 (Table B-1 MIL-Std 414) is used to determine the sample size. This sample may be further reduced to 12 when the standard deviation is known. There is no acceptance number since variables measurements are used in the determination of acceptance. It is possible for no values to be outside of specification in the sample and the lot to be rejected. Nevertheless, it is one of the conditions that must be tolerated since lot acceptability is based on the average and standard deviation (or range) of the lot relative to the specifications. When an upper specification limit is specified, the sequence of calculations in determining whether the lot is acceptable is as follows:

Table 9.16 Master Table for Normal and Tightened Inspection for Plans Based on Variable Unknown[a]

Sample size code letter	Sample size	Acceptable quality levels (normal inspection)													
		0.04	0.065	0.10	0.15	0.25	0.40	0.65	1.00	1.50	2.50	4.00	6.50	10.00	15.00
		k	*k*	*k*	*k*	*k*	*k*	*k*	*k*	*k*	*k*	*k*	*k*	*k*	*k*
B	3	↓	↓	↓	↓	↓	↓	↓	▼	▼	1.12	0.958	0.765	0.566	0.341
C	4	↓	↓	↓	↓	↓	↓	↓	1.45	1.34	1.17	1.01	0.814	0.617	0.393
D	5	↓	↓	↓	↓			1.65	1.53	1.40	1.24	1.07	0.874	0.675	0.455
E	7	↓	↓	↓	↓	2.00	1.88	1.75	1.62	1.50	1.33	1.15	0.955	0.755	0.536
F	10	↓	↓	↓	2.24	2.11	1.98	1.84	1.72	1.58	1.41	1.23	1.03	0.828	0.611
G	15	2.64	2.53	2.42	2.32	2.20	2.06	1.91	1.79	1.65	1.47	1.30	1.09	0.886	0.664
H	20	2.69	2.58	2.47	2.36	2.24	2.11	1.96	1.82	1.69	1.51	1.33	1.12	0.917	0.695
I	25	2.72	2.61	2.50	2.40	2.26	2.14	1.98	1.85	1.72	1.53	1.35	1.14	0.936	0.712
J	30	2.73	2.61	2.51	2.41	2.28	2.15	2.00	1.86	1.73	1.55	1.36	1.15	0.946	0.723
K	35	2.77	2.65	2.54	2.45	2.31	2.18	2.03	1.89	1.76	1.57	1.39	1.18	0.969	0.745
L	40	2.77	2.66	2.55	2.44	2.31	2.18	2.03	1.89	1.76	1.58	1.39	1.18	0.971	0.746
M	50	2.83	2.71	2.60	2.50	2.35	2.22	2.08	1.93	1.80	1.61	1.42	1.21	1.00	0.774
N	75	2.90	2.77	2.66	2.55	2.41	2.27	2.12	1.98	1.84	1.65	1.46	1.24	1.03	0.804
O	100	2.92	2.80	2.69	2.58	2.43	2.29	2.14	2.00	1.86	1.67	1.48	1.26	1.05	0.819
P	150	2.96	2.84	2.73	2.61	2.47	2.33	2.18	2.03	1.89	1.70	1.51	1.29	1.07	0.841
Q	200	2.97	2.85	2.73	2.62	2.47	2.33	2.18	2.04	1.89	1.70	1.51	1.29	1.07	0.845
		0.065	0.10	0.15	0.25	0.40	0.65	1.00	1.50	2.50	4.00	6.50	10.00	15.00	
		Acceptable quality levels (tightened inspection)													

Source: MIL–Std 414 (1957).

[a]All AQL values are in percent defective. use first sampling plan below arrow, that is, both sample size as well as *k* value. When sample size equals or exceeds lot size, every item in the lot must be inspected.

Step Information needed

1 Sample size: n
2 Sum of measurements: ΣX
3 Sum of squared measurements: ΣX^2
4 Correction factor (CF): $(\Sigma X)^2/n$
5 Corrected sum of squares (SS): $\Sigma X^2 - \text{CF}$
6 Variance (V): $\text{SS}/(n - 1)$
7 Estimate of lot standard deviations: V
8 Sample mean $\overline{X}$: $\Sigma X/n$
9 Specification limit (upper): U
10 The quantity: $(U - \overline{X})/s$
11 Acceptability constant: k (found in the tables)
12 Acceptability criterion: compare $(U - \overline{X})/s$ with k

The range (R), rather than the standard deviation, can also be used for determining lot acceptability. It must be kept in mind that the variables sampling plans can only be applied to a single characteristic at a time.

Summary

Knowing how to use sampling tables and when to apply them can be important in a world-class quality system. This appendix presents a brief overview. For a comprehensive treatment of sampling, see Schilling (1982).

References

American Society for Quality Control. *Definitions and Symbols for Acceptance Sampling By Attributes,* ASQC, Milwaukee, Wis.

ANSI/ASQC Z1.4-1981, formerly MIL-Std 105D, *Sampling Procedures and Tables for Inspection by Attributes,* American Society for Quality Control, Milwaukee, Wis.

ANSI/ASQC Z1.9-1980, formerly MIL-Std 414, *Sampling Procedures and Tables for Inspection by Variables for Percent Nonconforming,* American Society for Quality Control, Milwaukee, Wis.

Bowker, Albert H., and Henry P. Goods. *Sampling Inspection by Variables,* McGraw-Hill, New York.

Deming, W. Edwards (1987). *Out of the Crisis,* MIT Center for Advanced Engineering Study, Cambridge, Mass.

Dodge, Harold F., and Harry G. Romig (1959). *Sampling Inspection Tables,* 2nd ed., Wiley, New York.

Duncan, Acheson J. (1974). *Quality Control and Industrial Statistics,* 4th ed., Richard D. Irwin, Homewood, Ill.

Grant, Eugene L., and Richard S. Leavenworth (1980). *Statistical Quality Control,* 5th ed., McGraw-Hill, New York, Appendix 4.

Harris, D. H., and F. B. Chaney (1969). *Human Factors in Quality Assurance, New York.*

Military Handbook H53, *Guide for Sampling Inspection,* Office of the Assistant Secretary of Defense (Installations and Logistics), Washington, D.C. June 30, 1965.

MIL-Std 1235A, *Multilevel Continuous Sampling.*

Schilling, Edward G. (1982). *Acceptance Sampling in Quality Control,* Marcel Dekker, New York.

10 · Beyond Statistical Process Control

Introduction

When Bell Laboratories developed the applications of statistical methods to manufacturing processes under the guidance of Walter Shewhart (1983), one of the driving forces behind this evolution was the need of the company to maintain its product during field usage. Studies showed that costs associated with field repairs were often 10 times the cost of preventing defects in the first place. When laying telephone cables under the ocean, the cost of repair was certainly prohibitive. Few other companies had such lifetime warranties and so lacked the commitment to try to produce defect-free product. In fact, many companies often made substantial profit on their service operations, so there was even less motivation to strive for perfection.

In today's competitive international market, users expect products to operate when purchased and continue to operate in a maintenance-free manner for years to come. Companies whose products meet other consumer needs in performance, style, price, and reliability (requiring little or no maintenance) will outperform their competition. The best way to achieve this happy state is through integration of good design and manufacturing practices. Companies must produce products as close to the target value of the specifications as possible with as little variability as possible. Shown in Fig. 10.1 are five possible performance distributions with no defects being produced while cost or performance vary in each.

For example, in Fig. 10.1a any shift in the average and some defective product will start to be produced. Furthermore, if we assume the specification values to be realistic, the product performing at the extremes may drift out of acceptable performance limits during operation, causing service or warranty claims. This type of production is characterized by a fair amount of inspection to keep the process operating within specification. In Figure 10.1b there is room for process shift, but this type of process still needs some inspection. In Fig. 10.1c and d the processes are off-center and a shift toward the specification limits can cause out-of-tolerance product. If the items produced are to be incorporated into a higher-level assembly, the assembled unit may have more marginal or out-of-tolerance units. Careful control, requiring more people or added operations, adds cost. In Figure 10.1e, with the process center and the process capability much better than the specification limits, consideration can be given to eliminating control charts. Product performance such as this can be used to outperform competition and, depending on other factors, either lower prices or advertise superior performance and raise prices. This condition occurs most often when statistical methods are used for process control and continued efforts toward process optimization are applied.

Note that although it may be desirable to use statistical process control on all processes, there are practical limitations. Thus when a process exhibits performance that is much better than the specifications require and it is in statistical control, reducing the frequency of evaluation or eliminating the process control is probably worthwhile.

In a more formal sense, all these processes are capable of meeting specifications but can be differentiated by comparing the ratios of specification limits to process spread. A ratio of 1:1 means that the process capability can just meet the specification. Mathematically, this is the tolerance divided by the process capability 6σ and is called C_p. A ratio greater than 1:1 means that the process is performing better than the specification. When it reaches 3:1 or more, action should be taken to reduce costs by eliminating some inspection and supervision, tightening specifications, and perhaps increased prices. If the ratio is less than 1:l, the process is not capable of meeting specifications and is much more costly because it requires 100% test or inspection to remove those

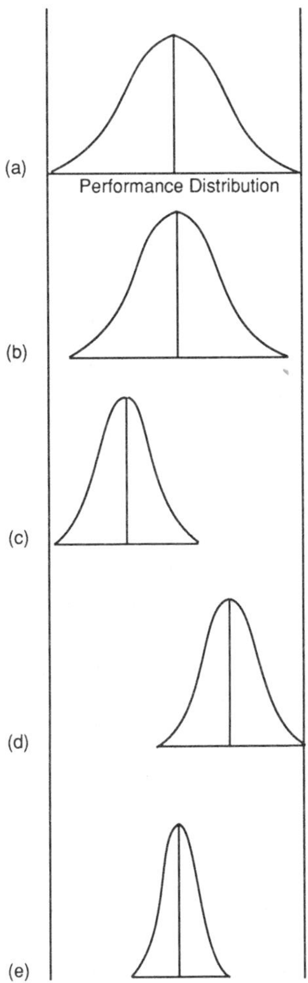

Figure 10.1. Performance distributions for conforming product. Frequency of occurrence: (a) Process is centered on specification target. Process capability equals specification limits. (b) Process is centered on specification target. Process capability is better than specification limits. (c) Process is off center on the low side of specification and process capability is better than specification limits. (d) Process is off center on the high side of specification and process capability is better than specification. (e) Process is centered on specification target and process capability is much better than specification limits.

items outside specification limits. Another important term in process capability is the ratio of the distance between the process mean and the nearest specification limit to one-half the process capability, or 3σ. This value should exceed 1 to be satisfactory. Do you or your technical and management team know what the process capabilities are? Is it worth knowing? Only if you want to beat the competition!

The Role of Statistics

This is not a book on statistics. But since statistics are one of the best ways to optimize processes, reduce the time from concept to market, and improve process capabilities, a brief overview of statistical applications is warranted.

The Pareto Distribution

The Pareto distribution first came to prominence in quality control in Juran's text (Juran, 1965). It is a convenient way to display attribute data in order to separate the dominant defects from the less frequent defects. It is a tool often used by quality circles to focus on problems. The display is a ranked bar graph, with the height of each bar proportional to the occurrence of the nonconformance. The order of the bars ranges from the most to the least frequent occurrence. The actual height of the bar can be the number of defects, percent defects, or defects as a percent of the total defects, as shown in Fig. 10.2a, b, and c respectively. Thus the principal defects can be addressed on a priority basis. When there are many possible causes for an outcome, it is common to find that a relatively few causes contribute a major effect. Thus three or four causes often contribute 60 to 90% of the problems.

This circumstance has also been called the 80/20 "rule," when 80% of the result is due to 20% of the causes. These numbers are really just a rough approximation and should not be regarded as gospel. This is true for rejects, sales distribution, inventories, costs, taxes, and more. It is a good idea to keep this rule in mind when considering data of any sort. The average is often a misleading statistic because it conveys the concept of uniformity when that is, indeed, not the case. Actually, the Pareto distribution can be replaced with a properly structured defect report that arranges defects in columns. This type of report can be manually generated as shown in Fig. 10.3a, or computer generated as shown in Fig. 10.3b.

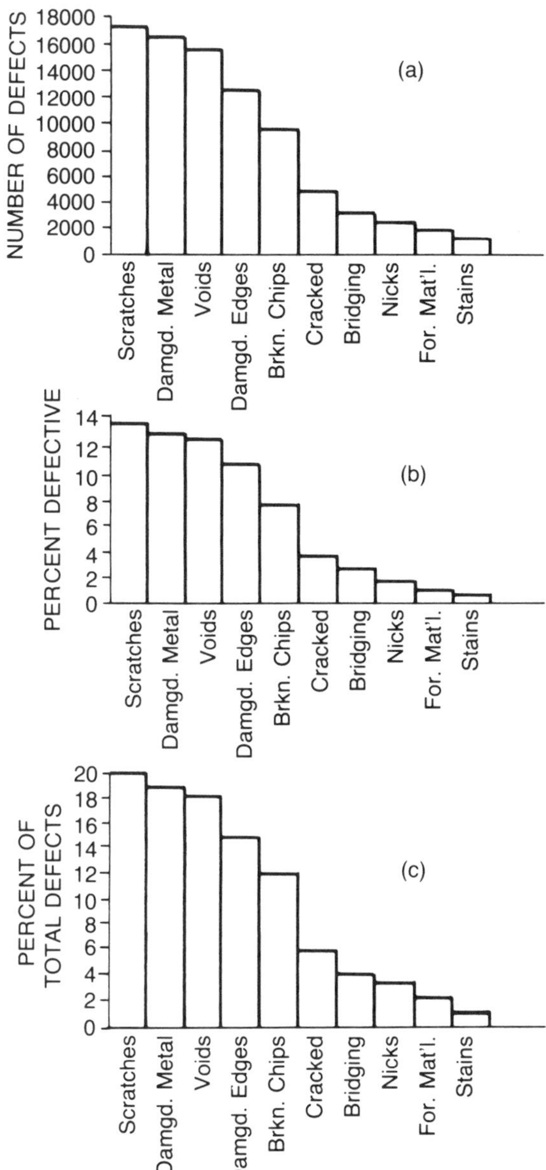

Figure 10.2 Pareto distributions.

				DEFECT TYPE																			
				Major Classif.					Major Classif.					Major Classif.					Major Classif.				
Date	No Eval.	No Good	No Def.	Defect A	Defect B	Defect C	Defect D	Defect E	Defect F	Defect G	Defect H	Defect I	Defect J	Defect K	Defect L	Defect M	Defect N	Defect O	Defect P	Defect Q	Defect R	Defect S	Defect T
XX	1026	1002	24	16					8														
XA	925	924	41	9	1				20			10			1								
XB	1027	985	42	18					14		5					5							
XC	1948	1126	22	12					10														
XD	1921	1827	94	37					32	4			6	6			3	1	5				
XE	986	909	77	41			3	2	27		4												
XF	1422	1400	22	6					16														
XG	1227	1201	26	3		1			20		1		1										

(a)

INTERNAL INSPECTION REPORT PAGE
DIVISION
PROJECT: DIVISION
04/01/85

FROM: 850325
TO: 850331

REPORTING STATION	# OF INSP	# OF REJ	PREJ	ORG	RESB	SOURCE ECN	RWRK	DEF	DPNI	DEFECT TYPE ASSY	CIRC	COMP	SUBS	COAT	DOCU	WIRE	SOLR	MECH	TEST	PACA	SOLD	TERM	BRAZ	WELD	BLAN
DCAS BRD INS	13	2	15.3	8	0	1	4	4	0.3	2	0	0	0	2	0	0	0	0	0	0	0	0	0	0	0
MACHINING	1052	92	8.7	1000	6	18	28	112	0.1	0	0	0	0	22	3	0	0	64	0	0	1	22	0	0	0
BOARD ASSY	97	15	15.4	73	19	1	4	57	0.6	13	2	5	0	3	9	0	0	0	0	0	25	0	0	0	0
PC FAB	2	0	0.0	2	0	0	0	0	0.0	0	0	0	0	0	0	0	0	0	0	0	0	0	0	0	0
MICROCIRCUIT	190	10	5.2	171	2	0	17	26	0.1	3	0	6	0	0	10	3	0	0	0	0	0	3	0	0	0
EMMA	87	7	8.0	75	2	0	10	54	0.6	0	0	4	0	0	48	0	0	2	0	0	0	0	0	0	0
WELD AREA	49	5	10.2	40	9	0	0	57	1.2	2	2	1	0	0	6	0	0	0	0	0	46	0	0	0	0
CONFORM COAT	151	10	6.6	141	4	0	6	11	0.1	1	0	2	0	1	6	1	0	0	0	0	0	0	0	0	0
SHUTTLE AREA	11	2	18.1	3	0	1	7	2	0.2	2	0	0	0	0	0	0	0	0	0	0	0	0	0	0	0
401 TEST MON	1	3	3.2	69	1	0	21	7	0.1	1	0	0	0	0	6	0	0	0	0	0	0	0	0	0	0
VIB TEST MON	11	1	9.0	4	0	0	7	1	0.1	0	0	0	0	1	0	0	0	0	0	0	0	0	0	0	0
T/V TEST MON	1	0	0.0	1	0	0	0	0	0.0	0	0	0	0	0	0	0	0	0	0	0	0	0	0	0	0
TOTALS:	1755	147	8.3	1587	43	21	104	331	0.2	24	4	18	0	29	88	4	0	66	0	0	72	25	0	0	0

(b)

Figure 10.3 (a) Manually generated defect report. Defects arranged in columns to permit easy identification of major defects. (b) Computer generated defect report. High defect occurrence shows in totals.

The Frequency Distribution or Histogram

A useful tool to show the behavior pattern of a product or process is the frequency distribution. This data display technique is used for measurements or variables data. By using cells and recording the frequency of occurrence at the value of that occurrence in each cell, a great deal more information is apparent than when the same data are listed in tabular form. For example, one can get an idea of the average, the spread, and the location of the distribution with regard to the specification limits and the shape of the distribution, provided that only 15 or so classifications or cells are used. It is also a convenient form to calculate the average and standard deviation. This insight is simply not present when data are examined in columns (see Fig. 10.4). A histo-

	.052	.048	.048	.045	.046
	.065	.053	.031	.041	.042
	.046	.038	.054	.042	.052
	.055	.049	.035	.055	.065
	.044	.042	.043	.065	.042
	.045	.065	.036	.046	.051
	.041	.037	.046	.053	.047
	.052	.040	.043	.044	.065
	.042	.045	.036	.035	.039
(a)	.045	.053	.047	.047	.05

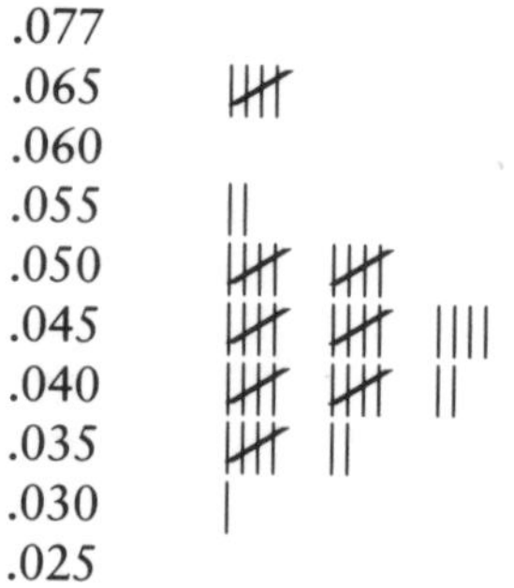

	.077	
	.065	卌
	.060	
	.055	||
	.050	卌 卌
	.045	卌 卌 ||||
	.040	卌 卌 ||
	.035	卌 ||
	.030	|
(b)	.025	

Figure 10.4 Tabulated readings. (b) Frequency distribution. The grouping at 0.065 is masked in the tabulated data, but it may be important. (c) Histogram of data shown in (a).

gram is a bar chart representation of these data. Note that there are no spaces between the bars except when there are zero readings at that value. A variation on the frequency distribution is a stem-and-leaf plot developed to enable retrieval of the actual data, as shown in Fig. 10.5. The simplicity and ease of the frequency distribution provided most of the information needed.

Control Charts

One limitation of the frequency distribution is that it neglects the effect of time. This is overcome by the use of the control chart. This is the principal tool in statistical process control and it has been discussed more extensively in Chapter 7. The control chart for variables, normally an $\overline{X}$ and range, $\overline{R}$, or an $\overline{X}$ and sigma, σ, chart, is used as a process control tool to identify process shifts. It should also be used for process improvement, which equates to reduction of the range or standard deviation. Continued reduction of the range and centering of the process may enable the identification of the process parameters that affect the process performance and lead to the elimination of the need for the control chart because key process parameters are automatically

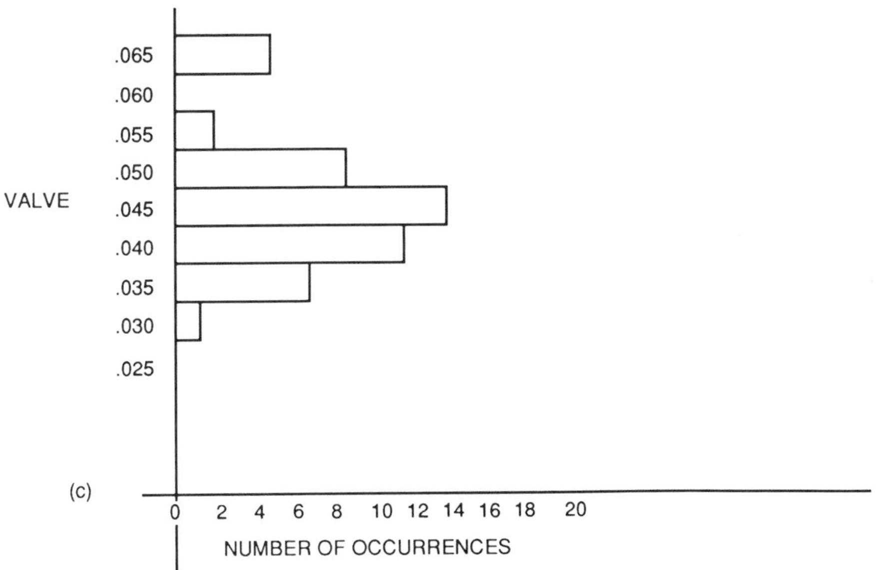

.06	5 5 5 5 5
.06	
.05	5 5
.05	2 2 3 3 4 3 2 1 0
.04	6 5 5 8 9 5 8 6 7 7 6 5 6 7
.04	4 2 1 2 0 3 3 4 2 1 2 1
.03	8 7 5 6 6 5 9
.03	1
.02	
.01	

Figure 10.5 Stem and leaf plot of data from Fig. 10.4a.

maintained within required limits. This will help improve quality and reduce costs. There are some other variations on control charts such as median and range charts, CuSum charts, narrow limit gauging, box and whisker charts, and others which are helpful tools in unique circumstances. The control charts for attributes, *p*, *np*, *c*, and *u* charts, differ in use from variables charts in that the objective is not to keep the process in control but to achieve breakthrough instead. Breakthrough, a term first coined for quality control use by Juran, is what helps to improve the process on a continuing basis. These improvements are possible as defect causes are identified and eliminated.

How is Breakthrough Achieved?

It is one thing to talk about breakthrough and quite another to achieve it. The most effective way I have found to achieve breakthrough is through the use of multifunctional teams whose charter is to improve operations. These teams must have a questioning attitude, operate in an environment that fosters change (as created by upper management), and be given the resources in terms of people, time, and perhaps, finances to explore improvements. Capital investment must be justified on the basis of payback, but the justification should include reduction in troubleshooting, rework, inspection, scrap, and repair as straightforward productivity increases. These teams can also function in system improvement areas, not just quality or productivity improvements, although system improvements usually connotate productivity improvements. Some methods used involve new technology, quality circle activity, quality improvement or corrective action programs, new tools or equipment, new processes, materials or parts, new designs, and others. One of the most powerful and cost-effective tools, often requiring no capital investment, is the tool of statistical design and analysis of experiments.

Statistical Design of Experiments

Up to this point, we have discussed techniques used in statistical process control. Now we go beyond SPC into new and very productive areas of experimental design and analysis using statistical methods. This book cannot begin to provide the details necessary for the reader to experiment with these methods. Instead, an appreciation for their use will be developed.

Experimental design utilizes combinations of factors operating at levels selected by the experimenters. By combining factors, a much more efficient experiment can be conducted than the traditional experimental approach of varying one factor and keeping the others constant. This enables multiple factors to be evaluated simultaneously along with their interaction effects (combination of factors) and can be accomplished with fewer experimental runs than are possible with traditional experimental methods.

There are many forms of such experimentation, but the simplest is the factorial experiment using two factors at two levels or a 2^2 experimental design. Thus an experimental layout might appear as shown in Fig. 10.6. Three factors at two levels, or a 2^3 design, would appear as the design layout shown in Fig. 10.7. The three factors might be temperature of solution, time of exposure, and solution concentration. One-half of a 2^3 design is termed a Latin square and the three factors are shown in Fig. 10.8. It is not uncommon to run designs consisting of five factors at two levels or a 2^5 (Fig. 10.9) which is 32 experimental combinations. There is a portion of this type of design that can also be run, called a fractional factorial. Thus a $1/4$ times 2^5 fractional factorial (Fig. 10.10) requires 8 combinations ($1/4 \times 32 = 8$) and can enable a comparison of all the main effects of the five factors, confounded with interactions. So with eight trials, five factors can be analyzed—certainly efficiency greater than the traditional one-at-a-time experimental process. The main effects are, however, mixed with interaction effects or confounded. So for this type of experimental design to be effective, one must have good reason to believe that combinations of variables are not likely to occur. Another representation for this design layout is:

						Interactions								
Run Number	A	B	C	D	E	AB	AC	AD	AE	BC	BD	BE	CD	DE
1	−	−	−	+	+	+	+	−	−	+	−	−	−	−
2	+	−	−	−	−	−	−	−	−	+	+	+	+	+
3	−	+	−	−	+	−	+	+	−	−	−	+	+	−
4	+	+	−	+	−	+	−	+	−	−	+	−	−	+
5	−	−	+	+	−	+	−	−	+	−	−	+	+	−
6	+	−	+	−	+	−	+	−	+	−	+	−	−	+
7	−	+	+	−	−	+	+	+	+	+	+	+	+	+

Where (+) values represent higher value and (−) values represent lower values. Under these circumstances, to evaluate the effect of A, runs 1, 3, 5, and 7 are averaged and compared with the average for

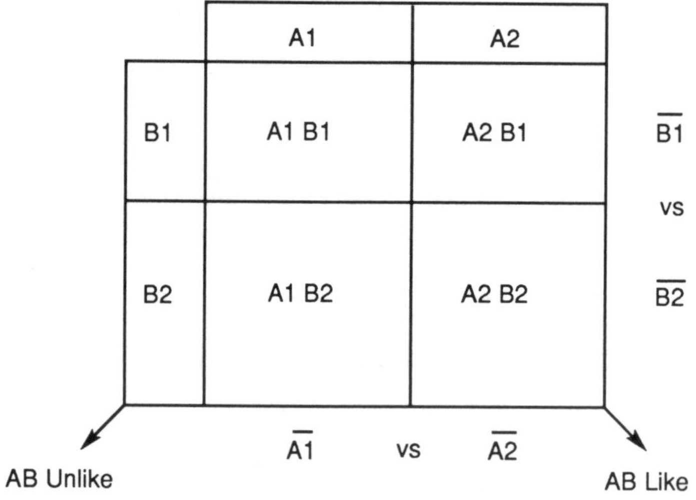

Figure 10.6 A 2 × 2 or 2^2 factorial design layout.

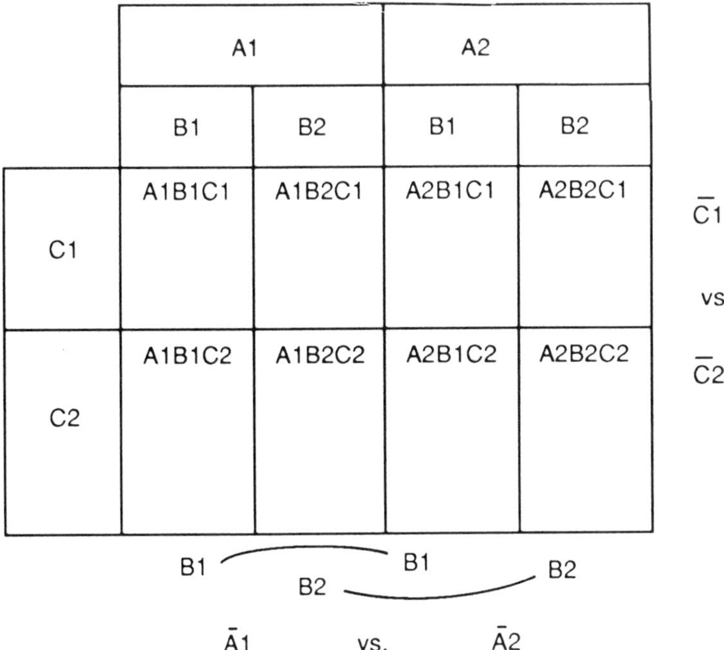

Figure 10.7 A 2^3 factorial design layout.

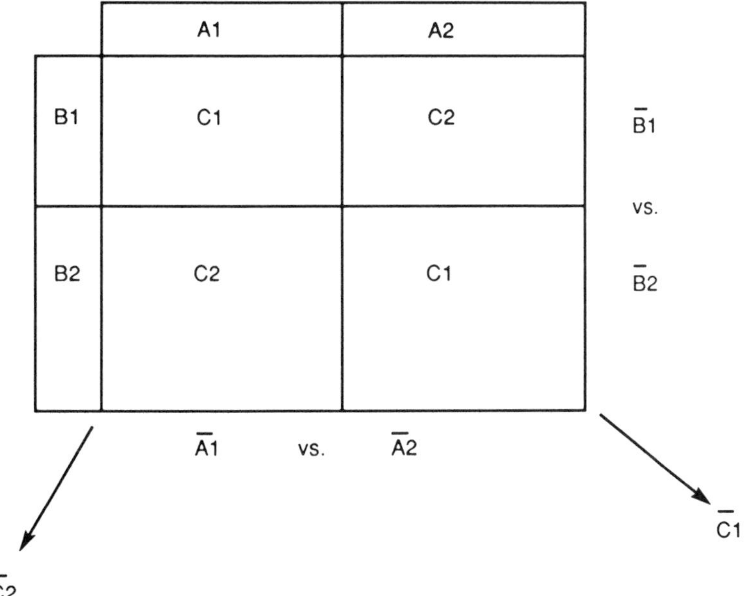

Figure 10.8 A Latin square design enables three process variables to be compared in four different experimental trial.

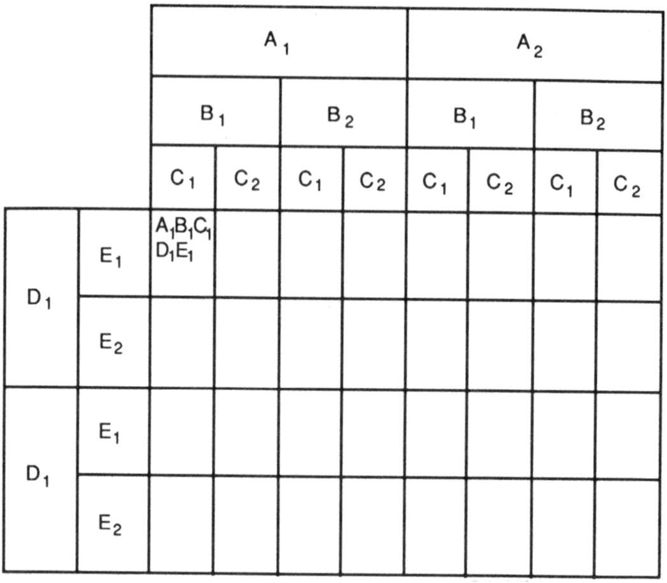

		A_1				A_2			
		B_1		B_2		B_1		B_2	
		C_1	C_2	C_1	C_2	C_1	C_2	C_1	C_2
D_1	E_1	$A_1B_1C_1$ D_1E_1							
	E_2								
D_1	E_1								
	E_2								

Figure 10.9 A 2^5 factorial design scheme. *A, B, C, D,* and *E* are process variables: 1 is the higher value, 2 is the lower value. Each of the 32 boxes is an experiment made of of high and low values of the process variable. The first block combines all high values, as shown.

runs 2, 4, 6, and 8. For B, runs 1, 2, 5, and 6 are averaged and compared with the average for runs 3, 4, 7, and 8. This is done for all five factors. The combination of main effects to provide interactions is shown in columns with two letters at the heading. By examining the signs in these columns, it becomes evident that the following confounding occurs.

Main effect of A confounded with BD and CE
Main effect of B confounded with AD
Main effect of C confounded with AE
Main effect of D confounded with AB
Main effect of E confounded with AC

To eliminate the confounding effect, another eight experiments arranged in a "foldover" design is necessary. This design merely reverses the original signs or repeats the experiments with high and low values interchanged. The eight runs would then appear as follows:

Run Number	A	B	C	D	E
1	+	+	+	−	−
2	−	+	+	+	+
3	+	−	+	+	−
4	−	−	+	−	+
5	+	+	−	−	+
6	−	+	−	+	−
7	+	−	−	+	+
8	−	−	−	−	−

Other types of designs, such as 2 × 3 (see Fig. 10.11) or 3 × 4 (see Fig. 10.12), and experimental schemes are available. These are discussed by Ott (1975) and Box et al. (1978). The power of the analytical method stems from the fact that rows and columns of data can be combined to have the effect of each row and column heading (variable selected), thus merging data from individual blocks.

Whenever data are gathered there will be differences among the results from different experiments. Statistical analysis enables these data to be analyzed to detect differences that are significant (in a statistical sense) from results that are not significant. Tests, such as analysis of variance or analysis of means, enable this discrimination to be determined at preassigned levels of confidence. Thus an analysis can provide 95 or 99% confidence that the difference shown is real and not simply due to random effects. This provides assurance that repetition of the test will show similar results. Hence process changes that improve performance should be formally incorporated only after verification by statistical design and analyses of experimental results. An alternative to this process is to graph the results with no preassigned confidence level and decide on an action plan based on the magnitude of the differences and other factors such as cost/benefit, technical analyses, competition, or some other relevant item. It is best to plan such experiments with a team consisting of technical personnel who are knowledgeable about the process together with those who are statistically astute. Thus the experiment will incorporate factors likely to affect results and the experimental design will be efficiently established so as to facilitate its conduct and its analysis.

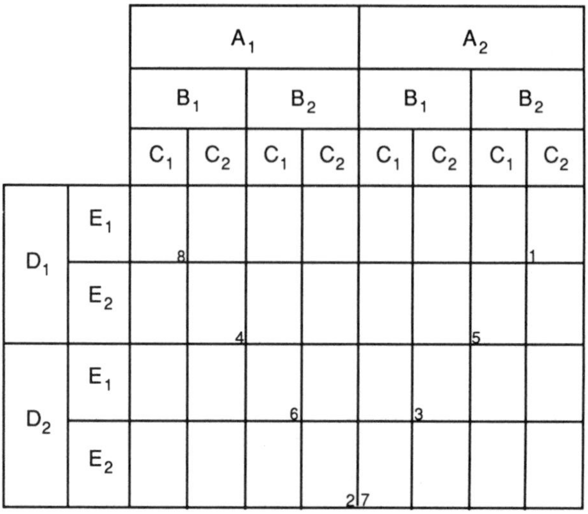

		A_1				A_2			
		B_1		B_2		B_1		B_2	
		C_1	C_2	C_1	C_2	C_1	C_2	C_1	C_2
D_1	E_1	8							1
	E_2		4					5	
D_2	E_1			6			3		
	E_2				2	7			

Figure 10.10 A 1/4 × 2^5 fractional factorial design scheme (also designed 2^{5-2}_{III}. *A, B, C, D,* and *E* are process variables: 1 is the higher value, 1 is the lower value. Each of the 32 boxes is an experiment made up of a combination of high and low values of the process variables. Numbers in the blocks correspond to the run numbers in Table 10.1.

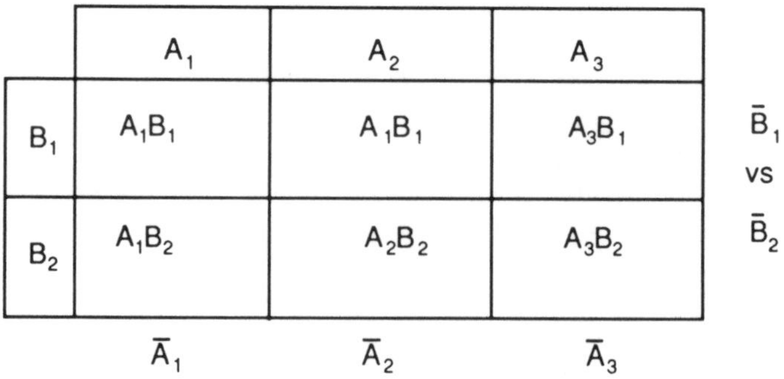

	A_1	A_2	A_3	
B_1	A_1B_1	A_1B_1	A_3B_1	$\bar{B}_1$ vs
B_2	A_1B_2	A_2B_2	A_3B_2	$\bar{B}_2$
	$\bar{A}_1$	$\bar{A}_2$	$\bar{A}_3$	

Figure 10.11 A 2 × 3 experimental design.

	A_1	A_2	A_3	
B_1	A_1B_1	A_2B_1	A_3B_1	$\bar{B}_1$
B_2	A_1B_2	A_2B_2	A_3B_2	$\bar{B}_2$
B_3	A_1B_3	A_2B_3	A_3B_3	$\bar{B}_3$
B_4	A_1B_4	A_2B_4	A_3B_4	$\bar{B}_4$
	$\bar{A}_1$	$\bar{A}_2$	$\bar{A}_3$	

Figure 10.12 A 3 × 4 experimental design.

One might ask why this is all necessary when the process parameters have been established by a competent group of scientists or engineers. There are at least two answers to this. One answer is that things change, things like material compositions, impurities in chemicals, central values of product distributions due to process shifts during the product manufacture, new vendors, equipment age or replacement, environmental factors, and so on. The second reason is that it is rare for a process to be optimally specified during development or even on the actual assembly line. Invariably, experimentation can improve process yields. If the process requires checking to verify performance because its central value (average) or spread (standard deviation) is too close to the tolerance, continued experimentation should take place.

These experimental procedures work with attribute data as well as with variables data, but more data are needed for attributes experimentation than for variables. This should pose few problems, however, for those dedicated to continuing process improvement.

One difficulty created in using statistical designs is that the ability to run the tests in a factory environment is often limited. Although it may be practical to keep the number of variables to no more than five, professional associates indicate that about eight variables can be handled efficiently. The practical ability to control the experiment and keep the experimental procedure randomized (to a reasonable level) without mixing up the parts or erring in the combination of parameters can be a major problem. The factory must be kept operational to produce salable product during all these experimental programs. Someone must be assigned to follow the experiment to make sure that it is performed as it was intended. The key point is that running these experiments is essential to gaining a leadership position among competitors.

Another difficulty in running experiments is interference with the production process. If off-line experimentation can be run, the variables being evaluated can be selected from a fairly wide range. This will provide greater assurance that the effect of that variable will be measured. Some processes, however, cannot be run off line, so it is necessary to experiment with the actual process. To do this, a technique known as evolutionary operation, developed by Box and Hunter (1978), can accommodate this situation. The experimental process uses small increments of change in the processing parameters so that the parametric values are operated within specified limits but at different bands within the specification. Although more data are required to evaluate the effects, the results can lead to process optimization, while maintaining production.

Computer Applications for Experimental Design

More and more of these methods are becoming computer-based for analysis, so less skill is needed and results are more quickly obtained. Care must be exercised, however, to assure the integrity of the data and to evaluate results graphically to ensure peculiarities in the data do not lead to incorrect conclusions. It may also be a good idea, depending on external factors, to repeat the experiment to make certain the results show up the same way before changes are introduced.

The Taguchi Method

A word must be said about the work being done by Genichi Taguchi, a Japanese engineer whose techniques are being applied in many major industrial organizations.

The design process advocated by Taguchi is one intended to develop robust designs, which will be stable over a wide range of external and internal variation. Factors called "noise factors" by Taguchi tend to cause variation during use and can lead to reduced performance and ultimately to lower customer satisfaction. By selecting product parameters using statistical experimental methods, product performance can be optimized to desensitize these parameters so they will shift less when exposed to noise factors.

Some statisticians dispute the validity of these approaches and neither mathematical nor statistical interpretation supports these methods. Nevertheless, the technique seems to work and it is being broadly applied in industry. One possible reason for its success is that effects of some parameters are large enough that statistical evaluations need not be extremely precise to realize the impact of the change.

Along with design of experiment methods, this technique can also be performed during the design phase. This enables more parametric values to be evaluated, affording a better opportunity to optimize the design for manufacturing use. As the ability to simulate performance increases, these evaluations can be done using computer analysis rather than actually making products. When a design is robust, it is more likely that manufacturing variability will not cause rejects and the field performance will be more stable when subjected to external factors such as heat, humidity, and time.

Another feature of Taguchi's approach is identification of the loss factor, in which he maintains that products whose performance is further away from nominal (or target) value causes the user to suffer a loss of performance that can be correlated with economic loss. There is also some controversy over this logic, but it is evident that products with smaller variation will be more satisfactory in use than products with wide variation. In effect, continual reduction in the process spread offers a tremendous competitive advantage. For further information on this method, see Taguchi (1986, 1987).

Using Experimental Design to Improve Product and Profit

A final problem associated with developing a statistical design process is management leadership and support. As in all other business matters, management must be supportive and in this situation must demonstrate leadership. There is an investment in time and money required for the experimental process, but the return has almost always proven to outweigh the investment. Unless there is commitment to continuing improvement, this process will not be sustained. This is a key ingredient in helping to establish world-class competition. It is what will enable reject rates to get to parts per million or parts per billion rather than fractions of a percent. This can be accomplished economically, so products will end up cheaper as well as better and will provide greater value to the user.

Case History: A Process Improvement Program

An example of the impact of the use of these methods occurred in the manufacture of nickel–cadmium batteries. When process improvements are desired, the first problem is to decide which factors to use and which levels to run. In this situation the research people were asked to identify all the known variables in the manufacturing process. Some 85 variables were identified. The list was circulated to quality engineering, manufacturing engineering, production, and design people, requesting any additions. After several weeks, 96 parameters had been identified. At a group meeting, these parameters were ranked 1 through 4 from most important to least important. Six factors received top votes from everyone present and two of these factors were selected to run in a 2 × 3 experiment (two factors at three levels). Some felt that a 14×25 experiment should be selected, but the difficulty associated with running such an experiment under factory conditions resulted in a consensus to run a simpler experimental design.

Part of this experiment was to reuse chemicals which were theoretically depleted during the processing of the product and therefore were being discarded after each use. The other factor was concentration of the chemical. There was no noticeable performance degradation of the product until 16 cycles were run, but the chemical concentration showed that the existing process was best. To provide a safety factor, 10 uses were established as the new standard, thus reducing material purchases to 10% of the prior procurement, a $50,000 annual savings. Simultaneously, the problem of disposal of used chemicals was reduced dramatically.

The experimental process continued using other variables, resulting in individual cell capacity increasing and a resultant yield improvement from 8% rejection to 0% rejection. This allowed cell testing to be eliminated, enabling the transfer of five quality control testers into production. As the process continually improved, it was found that performance was better than twice that advertised.

Rather than raise prices, fewer plates were installed in the batteries. This resulted in a product about 25% better than specification, with further dramatic cost reductions. As a result, the entire battery design concept was changed from a prescribed number of plates to use only enough plates to meet the specification by a comfortable margin. This type of success story can and must be achieved throughout American industry. The most powerful tool to help achieve this is the statistically designed experiment.

Correlation and Regression

Other statistical methods are helpful in the troubleshooting and problem-solving process. The use of scatter plots is taught to quality circles and is a graphical representation of two factors to illustrate their interdependence. One such example could plot a product characteristic against a process parameter. For example, if the charging rate of a battery affected its capacity, the plot would look like Fig. 10.13a, and if there was no effect, it would appear as shown in Fig. 10.13e.

There are simple statistical ways of analyzing these data to determine with a certain confidence whether or not correlation exists. One such test is the Olmstead–Tukey quadrant sum test (see Ott, 1975). Other more complex statistics allow the determination of a correlation coefficient based on the degree of dependence of one variable upon another. Figures 10.13a through e illustrate various degrees of correlation.

In addition, a mathematical equation can be derived based on the data enabling performance predictions of one factor based on another. Thus a regression equation can be developed. This can be done for more than two factors, and a multiple regression can be developed estimating the relationships among three or more factors. This can lead to further process optimization techniques.

Other Statistical Methods

Other statistical techniques for helping to make decisions are available to provide greater assurance in the decision-making process. Do you have multiple machines turning out products? If so, there are differences among the machines. Are these differences of economic consequence, and if so, are they real or do they occur by chance?

Statistical tests of significance help make the decision with greater certainty than intuition. If there are multiple heads on a machine, multiple cavities in a mold, or multiple operators, all of which have differences, statistical methods help determine the real from the chance variability. Precise management action can then be taken. Furthermore, methods such as these are not complex, but intelligent applications applied on a continuing basis help ferret out problem areas so that causes can be isolated and eliminated. Statistical methods, as powerful and helpful as they are, can only partially help improve operations.

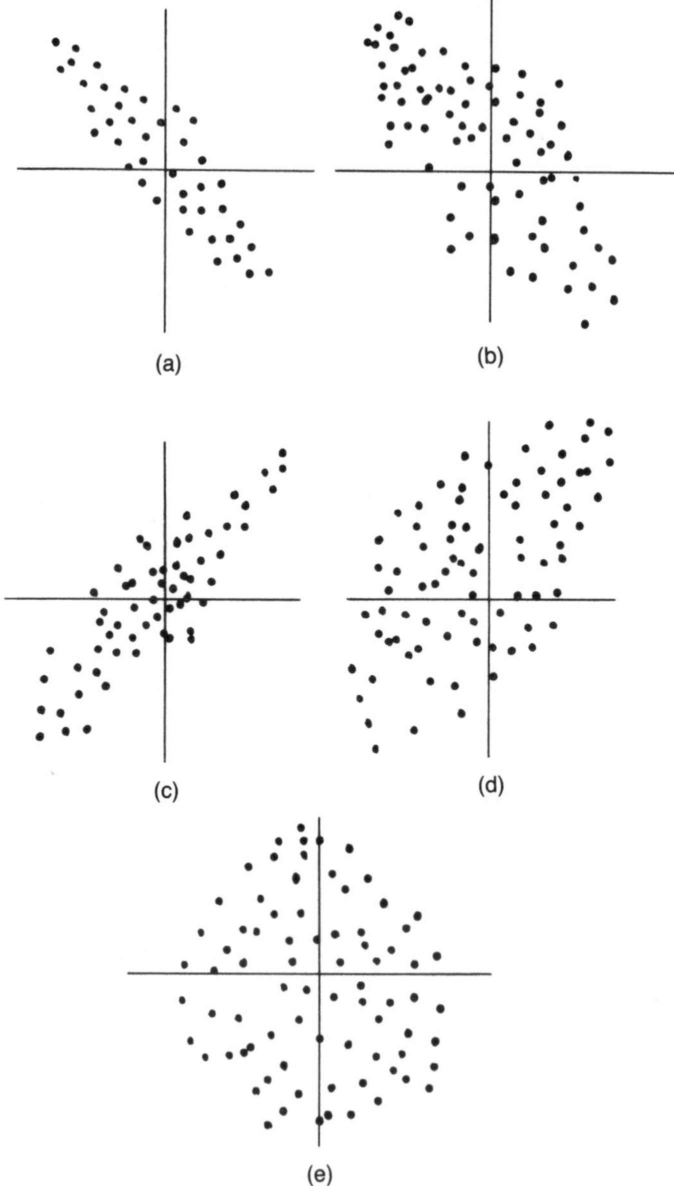

Figure 10.13 Sample correlation plots: scatter plots showing (a) good negative correlation, (b) fair negative correlation, (c) good positive correlation, (d) fair positive correlation, and (e) no correlation.

Industrial Engineering Techniques

Others, most notably the Japanese, keep evaluating processes and procedures in a never-ending drive to reduce costs and improve products. Fixing a process by creating a condition where errors cannot happen through process changes is a technique called Poka Yoke, advocated by Shigeo Shingo. In a simplified way, if a part is supposed to be inserted in a particular location and occasionally is omitted, Shingo installs a sensor so that if the part is inadvertently omitted, a bell rings and a light flashes or the conveyor line stops so that an operator can recognize the oversight and insert the part.

Another technique, known as *single-minute exchange of die* (SMED) is used by Shingo to reduce machine setup time. By studying the die insertion operations and using industrial engineering techniques, he is able to consistently reduce die setup time. He observed, for example, that new die and installation parts were not available at the time the machine was to be set up. Instead, dies and parts were searched for after the machine was shut down. Collecting and preparing the dies and parts during the machine operating time, he is able to reduce die setup time substantially. This was coupled with installation simplification by reducing the number of different parts used in the installation process and other well-known industrial engineering methods to develop the technique. It was given a catchy name, SMED, and institutionalized throughout industry in Japan. Application of this straightforward idea has helped to improve productivity and to reduce economic lot size runs because setup times were reduced from hours to minutes. A great deal of credit goes to Mr. Shingo for recognizing this time-wasting process and acting to correct it.

In a similar manner, Ryuji Fukuda developed and applied the *cause-and-effect diagram and cards* (CEDAC) discussed in Chapter 13. In essence it is a technique to make the process and the written procedure agree while striving for process simplification. Applying these methods throughout operations created many advantages. To remain competitive, companies must continue to seek out and eliminate time-wasting activities.

Summary

The purpose of this chapter is to broaden the perspective on the application of statistical methods to improve quality, yields, and productivity. There is much more to statistical methodology than statistical process control and considerably more than has been discussed in this chapter. It is necessary to be aware of these techniques if there is to be a genuine world-class quality environment in a company. A great many books and thousands of articles discussing methodology and application are available. Along with the application of industrial engineering techniques, and good management practices to create an environment for progress, statistical methods are vital to continued competitive success.

References

Besterfield, Dale H. (1986). *Quality Control*, 2nd ed., Prentice-Hall, Englewood Cliffs, N.J.

Box, G. E. P., W. S. Hunter, and J. S. Hunter, (1978). *Statistics for Experimenters*, Wiley, New York.

Braverman, J. D. (1981). *Fundamentals of Statistical Quality Control*, Prentice-Hall, Englewood Cliffs, N.J.

Burr, I. (1976). *Statistical Quality Control Methods*,

Caplan, Frank (1980). *The Quality System*, Chilton, Radnor, Pa.

Crosby, Philip (1979). *Quality Is Free*, McGraw-Hill, New York.

Crosby, Philip (1984). *Quality Without Tears*, McGraw-Hill, New York.

Davies, O. L. (1978). *The Design and Analysis of Industrial Experiment*, Longman, New York.

Deming W. Edwards (1987). *Out of the Crisis*, MIT Center for Advanced Engineering Study, Cambridge, Mass.

Deming, W. Edwards (1982). *Quality Productivity and Competitive Position*, MIT Center for Advanced Engineering Study, Cambridge, Mass.

Dixon, Wilfrid J. and Frank J. Massey, Jr. (1969). *Introduction to Statistical Analysis*, 3rd ed., McGraw-Hill, New York.

Duncan, Acheson J. (1974). *Quality Control and Industrial Statistics*, 4th ed., Richard D. Irwin, Homewood, Ill.

Fukuda, Fyuji (1983). *Managerial Engineering: Techniques for improving Quality and Productivity in the Work Place*, Productivity, Inc., Stamford, Conn.

Grant, E. L. and R. S. Leavenworth (1988). *Statistical Quality Control*, 6th ed. McGraw-Hill,

Ishikawa, K. (1982). *Guide to Quality Control*, Asian Productivity Organization.

Ishikawa, Kaoru and David J. Lu, translator (1985). *What Is Total Quality Control? The Japanese Way,* Prentice-Hall, Englewood Cliffs, N.J.

Juran, J. M. (1965). *Managerial Breakthrough,* McGraw-Hill, New York.

Juran, J. M. (1988). *Quality Control Handbook,* 4th ed., McGraw-Hill, New York.

Juran, J. M., and Gryna, F. (1970). *Quality Planning and Analysis,* McGraw-Hill, New York.

Lindgren, B. W., and G. McElrath (1969). *Introduction to Probability and Statistics,* 3rd ed., Macmillan, New York.

Moroney, M. J. (1968). *Facts from Figures,* Pelican Books, New York.

Ott, Ellis R. (1975). *Process Quality Control,* McGraw-Hill, New York.

Scherkenbach, W. W. (1987). *The Deming Route,* CFE Press,

Schilling, E. (1982). *Acceptance Sampling in Quality Control,* Marcel Dekker, New York.

Shewhart, W. A. (1983). *Economic Control of Quality of Manufactured Products,*

Shingo, Shigeo (1986). *Zero Quality Control: Source Inspection and the Poka-Yoke System,* Productivity Press, Cambridge, Mass.

Snee, R., L. Hare, and R. Trout (1985). *Experiments in Industry, Design, Analysis and Interpretation of Results,* Quality Press, ASQC, Milwaukee, Wis.

Taguchi, G. (1986). *Introduction to Quality Engineering,* Kraus-Thomson, Millwood, N.Y.

Taguchi, G. (1987). *The System of Experimental Design: Quality Resources,* Kraus-Thomson, Millwood, N.Y.

Taguchi, G., and Y. Wu (1980). *Introduction of Off-Line Quality Control,* Central Japan Quality Control Association; available from American Supplier Institute, Romulus, Mich.

Western Electric (1956). *Western Electric Handbook: Statistical Quality Control,* Western Electric, New York.

Wheeler, Donald J., and David S. Chambers (1986). *Understanding Statistical Process Control,* Statistical Process Controls, Inc., Knoxville, Tenn.

11 · Quality Control and High Technology

Introduction

In general, quality control methods, particularly statistical methods, are intended to enable continuous improvements in products, processes, and systems. The use of these approaches tends to increase yields by reducing product variability through better process controls and a more appropriate selection of design and processing parameters. Thus the apparent need for capital investment for new equipment and facilities may be postponed or eliminated entirely. For example, in one issue of *Boardroom* (see Godfrey, 1985), B. Godfrey, president of the Juran Institute, cites an instance in which a vinyl tile manufacturer was going to purchase a new furnace because tiles were warping during the curing process. After extensive study, engineering analysis indicated that a new oven with better heating controls could solve the problem. By using statistically designed experiments, it was found that the amount of lime in the clay could be increased, thus eliminating the warpage. Not only was the purchase of a new furnace unnecessary, but the cost of removing the extra lime was eliminated. In his book *Process Quality Control,* Ellis R. Ott (1975) provides many other examples of experimental designs in which analyses showed many situations with improper setting of process variables. Thus there are many opportunities for inexpensive process improvements which can be done quickly and provide both short- and long-term profit enhancement. A fertile field for finding opportunities lies in challenging the design and process specifications. Statistical techniques that evaluate alternatives often identify ways of optimizing performance and yield.

Sometimes, these improvements are just not possible without investing in new equipment, new materials, or new processes. Then it may just be good business to make the investment, because the product performance needs to be improved to maintain or gain a position of leadership. There are also times when product advances demand new capital investment. Process improvements using statistical methods can minimize the need for capital investment and provide substantial returns. These methods should not only be exploited when capital investment is involved but should be used continuously on a day-to-day basis to effect process and product improvements.

The objectives are to improve customer satisfaction through a superior product at a low-enough cost to capture market share and make a profit. Although the general theme of this book is better profits through higher yields and attendant lower costs, it is recognized that there are other ways to build customer loyalty. A new product fulfilling a need or desire in the marketplace can achieve success. The product cannot have inherent flaws or weaknesses presenting health and safety risks or inconvenience the user through excessive or costly repairs. Invariably, a useful product with consistently good performance and low maintenance will create a market. High technology helps build better products with lower costs for existing product as well as new products. High technology relates to products and processes as well as to services.

The Impact of High Technology

High technology relates to quality in several ways. One is the use of high-technology equipment in the conduct of the quality function for inspection, test, nondestructive testing, analysis, and use of proven computer-based standards. Another is through high-tech application in design, manufacturing, service, and systems with new or advanced processes, materials, machines and equipment, parts, and computer systems. Although the contribution of the labor force is neglected in this discussion, they play a significant role in the successful applications of high technology (see Chapter 13).

Processes

In electronic manufacture, soldering is one cause of rejects resulting in touch-up, rework, factory, and field failures. While the use of statistical methods can improve the selection of the many processing parameters, the application of statistical process control can help maintain a high-yield process. The use of a totally different technology, such as vapor-phase reflow soldering coupled with the use of surface-mounted devices, can improve the solder joint yields dramatically. This requires designers to use the surface-mounted devices in their designs, and the manufacturing department to perform the vapor-phase reflow operations and learn or develop high-yield processes for this application. The literature provides many examples of large improvement yields along with improved product features such as reduced product size, ease of repair, lower weight, and other improvements. There is no doubt that newer, less costly, better-yielding designs will emerge. World-class companies stay abreast of these developments and learn how to introduce them through astute integrated design and manufacturing processes.

The availability of microprocessors on equipment has enabled more precise controls to be installed on many processes. This provides for feedback of error signals to correct or stabilize processing variables, resulting in more consistent processing. To a degree, these controllers can augment or replace the use of control charts and keep processes behaving more uniformly, resulting in a more consistent product. A control chart would reflect this condition by showing a smaller range (or standard deviation) so that the 3σ values would be more tightly spread around the average.

In materials areas, such advances as higher-purity silicon have enabled the development of larger semiconductor chips, thereby placing more functions on a single chip improving the reliability and reducing the cost of assembly. This development, coupled with the capability to design customized large-scale integrated circuits (LSICA) or application-specific integrated circuits (ASICs), allows the creation of specific functional designs, greatly reducing parts counts and assembly costs while simultaneously improving reliability and quality. Further, the advent of high-speed digital communication enables a designer in one location of the country to create his or her own unique design and transfer data to a semiconductor manufacturer for fabrication in another part of the country (or the world). This trend will continue and result in shorter design cycles and lower costs as well. The advent of gallium arsenide can enable faster speeds and closer spacings on semiconductor chips, but at a higher cost. Developments in ceramics combine to create new opportunities for product improvements. Composite materials provide lighter weight and improved strengths. Statistical quality control can play a role in hastening these developments through the use of statistically designed experiments (see Chapter 10) and controlling the processes used to make these materials. These and a myriad of other developments must be of prime interest to companies striving for success in the marketplace. These advances will continue and must be viewed as major contributors to better product quality, shorter cycles between design and product release, and higher profitability.

In offices, the automation revolution is improving efficiency and accuracy in performing clerical, secretarial, and other nonfactory tasks. The speed, efficiency, and accuracy of converting sales to schedule changes and forecasts is continuing. The processing of purchase orders is moving toward computer-generated requisitions, purchase orders, and direct computer-to-computer purchases. Data storage on film, fiche, and optical disks reduces filing time and through automatic filing and retrieval systems improves the accuracy and speed of data retrieval and reduced storage space. Efforts in quality control are shifting to participation in systems development and systems evaluation to determine what new control systems are needed and working toward continual improvement in systems accuracy. New failure types are emerging and these must be addressed and corrected. For example, in word processing, words, phrases, paragraphs, or entire pages may be mislocated more easily than when using typewriters, while typographical and spelling errors are reduced. Relational data bases using common computer data base information for a variety of outputs must be structured to provide error-free reporting, data, and other output. The format of the data can enable better intelligence and make decision making easier and more accurate.

Improved processes and more uniform performance through the application of robotics also changes the role of the quality function. Greater effort must be expended in establishing the robotic capability to meet specification requirements and less effort diverted to inspection to achieve lower cost and higher-quality products. Instead of a great deal of inspection, process audits performed in the right amount can be used to control the process and assure compliant product. This has proven to be more efficient when applied in situations where process capability is well within specifications.

Also, relevant to the quality process are the developments in computer applications. In mechanical areas, for example, data bases used with computer-aided design (CAD) can be used to download geometric information to computer numerically controlled (CNC) machine tools. These, in conjunction with other computer programs, create a computer-aided manufacturing (CAM) system enabling the machinists to use their expertise in machining materials. These also enable the creation of programs combining the design expertise with manufacturing expertise to manufacture parts exactly as the design intended. An integrated computer-aided quality (CAQ) system can be used, if necessary, to verify parts integrity. On the other hand, yields may be so good that inspection is not necessary or can be performed during the machining process.

Simulation, coupled with the ability to perform stress, thermal, and other analyses of the design helps create more error-free designs, thereby reducing failures in the manufacturing process and in field use. The use of telecommunications enables remote sites to use the same data bases to manufacture a larger percentage of products in conformance with specifications. One example is General Motors' Manufacturing Automation Protocol (MAP), being installed in their own and supplier plants. Prime Computer Medusa software coupled with their Graphical Numerical Control Software similarly couples design data bases with manufacturing. This capability enables broadened use of the machine shop supplies to support manufacturing needs with better conformance, faster delivery, and potentially lower costs. This is only one step removed from approaches that will include expert systems in programming for machining parameters that will enable manufacturers to provide good parts more rapidly, thereby reducing the cycle from design to marketing. The reduction in time from concept to end product is a significant factor in beating the competition.

The quality system must be capable of functioning in the system design environment in order to recognize and prevent new types of problems. For example, in CAD/CAM when a product design change is introduced, the design model, not just the dimensional data in the data base, must be changed. If the latter occurs, the drawing, if one is used, will show one dimension while the machines will produce to the other in the model file. Thus the operational system must be changed and the CAD operators must be trained to do the job correctly.

Computer-aided engineering (CAE) systems enable engineers to play "what if" games with their designs to assure a more fault-tolerant design, one whose performance remains stable while experiencing greater processing and part variability. This robust design will result in easier manufacturing, fewer engineering changes and higher yields. If these yields are high enough, it enables the reduction or elimination of in-process inspection and tests. Ultimately, this results in greatly reduced cost and better product. These costs can be converted to profit to pay for improvements, provide greater returns, or enable price reductions to gain increased market share. In the long run, consumers benefit by getting greater value for their money.

Despite the advantages of CAE, the selection of parametric values for various parts does not necessarily evaluate product performance over a wide range of processing variables. The use of statistical methods to evaluate product stability assures that performance is constant despite parts, materials, or process variability. This will result in higher process yields and the need for less inspection and test. So statistical design of experiments can help select designs optimized for production and use environments.

Further developments along the lines of expert systems and artificial intelligence place knowledge into computer systems. Factory or office operations whose operating systems are incorporated into an expert system can potentially provide more rapid solutions to problems. It is necessary, however, to incorporate the quality engineering or management knowledge into the computer-based expert system. This way of building a corporate memory using past experience can help prevent future problems in new designs and rapidly resolve problems on current designs. As artificial intelligence comes of age and some reasoning power is built into software, the practical statistical tools can be incorporated to provide statistically sound and more robust solutions.

The interface of the quality function with other business progress also falls into the category of quality change as technology changes. The concept of just-in-time (JIT) delivery to support manufacturing needs while keeping inventory costs low has a major relationship to the quality function. For JIT to work, the product received must be defect-free so that it can be fed into the assembly operation upon receipt. Quality's role is to see to it that the suppliers have been properly selected and that the suppliers' manufacturing, quality system, and general management can and will support the effort necessary to provide products with virtually zero defects. In addition, the quality functions should reduce its own incoming inspection effort in consonance with the elimination of the incoming inspection (or test) activity. The JIT concept also applies to in-process work, so defect-free production is a necessity as well.

This drive for the reduction of inventory and the associated reduction in the cost of carrying this inventory should also penetrate in-process inventory. It must be recognized that the accumulation of rejected material awaiting analysis and repair increases work in-process inventory and stretches the production cycle. This forces purchasing personnel to provide parts sooner to allow for added delays and longer manufacturing cycles. Thus reducing inventory at receipt and in process enables lower inventory costs, enables more inventory turns, and reduces costs further.

Testing Philosophy

The aspect of reducing test and inspection warrants detailed commentary. The advanced concepts and capabilities of current automatic test equipment can make a major contribution to productivity growth. Strategic application of these capabilities can result in a substantial contribution to the competitive position of any organization. What is needed is a fresh look at how to develop and apply these strategies in today's environment.

In the first place, it must be recognized that testing is necessary only if defects are present in the unit under test. If there are no defects, why bother to test? Ideally, the product can be assembled completely and tested only at the end of the line. Why isn't this done more often? Why is it necessary to test at subassembly levels? The reason is evident: because design, parts, processes, and workmanship are not sufficiently free of defects to assure that the final product will pass all its requirements often enough to enable in-process testing to be eliminated. When a defect occurs at the final test station, the time and cost to resolve the deficiency are perceived to outweigh the cost of testing earlier in the assembly process. If the nonconformance reaches the consumer, costs are substantially greater and customer dissatisfaction will reduce market share.

In general, the earlier a problem is found, the easier it is to correct. Finding and correcting defective products is not the principal reason for testing. The main reason for testing early is to find the causes of problems and correct them. One of the secrets of the higher productivity realized in Japan and some other foreign countries is the substantial reduction of rejects. It is not uncommon, for example, to find an electronic product as complex as a television set to have only 1 reject in 1000 units produced. To achieve this type of result, it is essential to use the information generated by testing to eliminate the causative agents. When this is achieved, testing points can then be eliminated.

Let's take a closer look at some specifics. Assume that the earlier a defect is found, the cheaper it is to repair—certainly a logical assumption. Simulation can enable the detection of poor or marginal designs. Rejects due to design deficiencies will not result because these inadequacies can be eliminated before the design is committed to hardware. Some circuits can be tested for parametric values and frequency distributions can be made to ascertain the average and spread as compared to the specification limits. These analyses of pilot models or early production can be fed back to design and manufacturing for correction as necessary.

The next problem to be addressed is where to do the testing in the product flow. Although it is fairly obvious that the final products must be tested, it is not at all evident where testing at prior assembly levels should take place. It must be borne in mind that testing is not an end unto itself. Testing is done only because product performance is not known without it. The strategy must be twofold:

1. Testing during assembly should be done principally to generate information that can be used to correct conditions that caused the defect in the first place.
2. The testing points should be minimized in keeping with the findings at that test point. In other words, as test information is generated and used to correct problems, test points should be scrutinized to determine whether those test points can be eliminated.

To satisfy the statements above, it is necessary to do four things:

1. Analyze any deficiency promptly.
2. Summarize data to help focus on defect causes.
3. Take action to correct the cause.
4. Follow up to make sure that the corrective action worked and continues to work.

Modern automatic test equipment (ATE) enables prompt and accurate defect analysis due to the fault isolation and diagnostics capability built into the test system. There also exists capability in the data-logging feature of the ATE to provide summaries to be accumulated rapidly and in tabular or graphical form. A key ingredient in the data analysis is to provide information that is timely, accurate, and scannable, so the defects can be ranked and assigned to the appropriate individuals for investigation and corrective action, action that prevents recurrence.

If these elements are performed properly, the corrective action will result in decreasing rejections at the intermediate test points. When this occurs, it is mandatory to evaluate the need for testing at the particular point. An economic trade-off can be made by considering the cost of testing versus the additional cost of allowing a small percentage of defects to reach the next test level, where repair may be more costly (assuming that the deficiency will be detected). The benefits of operating in this matter may be demonstrated by examining the simplified flowcharts for an electronic assembly shown in Chapter 5.

When rejects are allowed to continue by not analyzing them promptly and correctly, they cause additional costs and reduced productivity. These added costs can be tangible as in the case of added scrap, rework, retest, reinspection, and larger inventories of material awaiting repair or intangible in the form of poor work attitudes or customer resistance to sales due to poor performance in the field. Despite the efficiency of modern ATE, products experiencing high rejections in house are likely to experience higher return rates. This results in higher warranty costs due to the fact that more marginal products either pass the test and subsequently fail in the field, or more good products are marginal and then fail in use. It is an undesirable and unwise management decision to allow such a state of affairs to continue.

Further consideration of the flowcharts shows that as corrective measures are introduced and defect rates drop, fewer test points are needed. This has the effect of shortening the manufacturing cycle reducing the lead time and work in process inventory and improving ability to meet schedules, as well as delivering more satisfactory products. If all these good things occur, the test equipment can be underutilized or less test equipment may be needed—at least in the short term. As costs decrease and product quality increases, either profit margins rise or price cuts can be passed along to the consumer, making products more competitive. In the long run, then, business can grow, and as needs rise this ATE can be used on more production or more product lines.

Production testing of electronic products generally falls into two categories: functional testing, in which the product is operated to determine its conformance to performance specifications, and in-circuit testing, in which individual part values are checked and short circuits, open circuits, reversed, wrong, or missing parts, and leakage paths are detected. Some newer equipment can perform both tests. Test technology will continue to improve to enable rapid problem detection. But this emphasis seems misplaced. Instead of working toward prevention of problems, great sums of money are spent to find the problem after it has occurred.

The new concepts in automatic testing utilize in-circuit testing on the premise that most functional failures are the result of poor workmanship, defective parts, or marginal processes. Although this may be true, one must still be concerned with function, and functional tests still need to be done. If, in fact, analyses of defects can be done quickly and accurately using the functional tester, one must question the need for in-circuit testing. Some justification can be seen due to lower programming costs for in-circuit testers. But if functional testing is still to be done, these are just added costs. A new industrial product, the in-circuit tester, has been created because we have been unable to reduce defects low enough to rely on functional testing alone. If functional testing only were performed, test equipment, test programs, jigs, and fixtures needed to perform in-circuit testing would be eliminated. Yet little effort has been expended to achieve this state of affairs. It is just easier to buy new in-circuit test equipment and incur all the associated costs rather than to design and build the product right the first time. This mentality leads to less efficient manufacturing and worsened competitive position. More testing adds to product cost. The key to productivity is to use rapid but minimum testing with accurate feedback and prompt, permanent corrective action regardless of the test system used.

New Technology in the Quality Function

New technology tools in the quality control function are growing and will continue to be developed. Improved measuring tools provide greater accuracy and quicker evaluation. Built-in microprocessors allow automated determination of statistical parameters such as the average and standard deviations and frequency distributions. They also simplify the application of statistical process controls. So the training of production operators in the use of the inspection tools is easier, but it is more important for these operators to have an understanding of SPC so that proper actions are taken when results are obtained. The focus then is to train operators in understanding what the measurement instruments are providing. Inspectors may be replaced by production operators. This has the positive effects of providing the actual producer with the wherewithal of evaluating operations under their control. They can correct problems and reduce the labor content.

The advent and continued improvement of machine vision systems can speed up the performance and increase the accuracy of inspection. They provide data in a more timely fashion and in a format that is easier to interpret and use. In their book, Harris and Chaney (1969) report on a study made on inspection efficiency, showing that the ability of an inspector to detect a defect when one is present varies with the complexity of the product, number of characteristics examined, and of course, inspector fatigue. These values vary from a low of 10% (finding 1 defect in 10) to a high of about 95%. The application of this equipment can greatly improve the inspection efficiency and reduce the cost of inspection, but even greater benefits can be realized by the use of this information for application to determine problem causes and eliminate them. It is shortsighted management to become dependent on the screening ability of these vision systems to remove nonconformities. It is usually much better to eliminate the problem cause and not have to have the nonconformance in the first place. Correction and prevention are the key ingredients. Further progress will continue to be made, enabling integration of computer-aided quality systems into CAD/CAM systems. The incorporation of measurement and feedback directly into manufacturing equipment to enable production of defect-free product has already started and will continue to expand. Sophisticated, nondestructive test equipment to help evaluate products and processes can isolate subtle shifts in material content or process variables. They provide information to engineers to enable problems to be isolated and eliminated. Services are available to companies that, in their drive for uniformly conforming products, do not have the continuing need nor the capital required for equipment to perform these analyses.

Summary

The application of new technology in the design, analysis, manufacture, and quality control of products affects the quality system used. In a general way, the quality effort will be more concentrated at the earlier phases of product development, although the specific direction the quality function takes is affected by the particular circumstances. A careful evaluation of these conditions is necessary to provide the right quality activity, statistical design of experiments, statistical process control, inspection, testing, and computer or robotic system to end up with a system providing low-cost, high-quality products that satisfy the marketplace. High technology will continue to provide equipment to enhance design, manufacture, and quality control of products and processes. Vigilance must be maintained to find and incorporate cost-effective improvements to assure continued performance at a world-class level.

References

Godfrey, B. (1985). "Quality Control, and More," *Boardroom Reports,* Vol. 14, No. 4, Boardroom Reports, Inc., New York.

Harris, D., and F. Chaney (1969). *Human Factors in Quality Assurance,* Wiley, New York.

Ott, Ellis R. (1975). *Process Quality Control,* McGraw-Hill, New York.

12 · Controlling Vendor-Supplied Items

Introduction

To produce a product or service that is competitive in the world marketplace, all avenues of cost contribution and hence potential cost avoidance must be examined. A major contributor to cost, often ranging between 40 and 70% of product cost, is associated with suppliers. It is imperative, therefore, to address costs and quality as they relate to vendor-supplied items. The total costs associated with these items include the product costs, the cost associated with incoming quality verification, inventory costs, and any costs associated with subsequent in-house rework, retest, and warranty losses due to vendor product.

Since vendor and subcontractors are part of the design and production teams, maximum utilization should be made of their expertise. There is a tendency on the part of a company's designers to ignore the contribution vendors can make to the design because of their experience and internal design capability. Normally, the procurement activity lacks the technical background to encourage this mode of operation, and frequently, the design group adopts a parochial attitude toward its own design. The quality function is in a good position to see that vendor expertise is incorporated into designs and tolerance setting.

Benefits of Good Supplier Quality

It has been the author's experience that good vendors can help reduce problems significantly by knowledge of their own process. Frequently, they can tell which tolerances are easy to meet and which are difficult. Not only can this result in higher levels of compliance, but it can also result in lower product costs. Vendors can also suggest feasible design or manufacturing alternatives. With regard to incoming product verification, these activities should be considered as a target for elimination. If the vendor is doing the job correctly, then the process controls, coupled with any necessary final test and inspection, should provide defect-free products. These products can then be shipped directly to inventory or the assembly floor for use, with no incoming product verification. This is the foundation of just-in-time inventory practice. The incoming quality function cannot be eliminated, though, until results have demonstrated that defect-free products can consistently be obtained from the suppliers. This may be a lengthy process, but the savings can be so substantial that it forces world-class competitors to strive for this goal.

Inventory costs can be reduced substantially by using vendor-supplied items shortly after receipt or directly upon receipt. This requires not only delivery of conforming product but a sophisticated production control operation that can develop appropriate schedules. Furthermore, manufacturing cycle time is reduced, contributing further to lower cost, and the cushion provided by inventory is eliminated, forcing data collected during the manufacturing process to be used rapidly in order to reduce nonconformities.

How to Get There

Getting defect-free products from suppliers is no simple matter. It requires a carefully developed program encompassing several activities: (1) assessing the capability of the supplier to meet the specifications, (2) supplier communications, and, (3) receiving and on-site quality control.

Assessing the Capability

Supplier capability can best be determined by a pre-award survey and past performance. The type of pre-award survey should reflect the type of supplies: vendor-manufactured items versus parts provided by a distributor; special items of the user's design versus vendor design, degree of mechanization, level of computerization, advanced technological items, and the supplier's process capabilities compared to specification requirements. For example, when surveying a supplier whose product requires advanced technology, an evaluation must be made of the vendor's technical competence. Design engineers from the procuring company should participate in the evaluation through interview and discussions with the suppliers designers. It is not sufficient merely to perform a quality survey. Prior to production, design reviews may be dictated to assure conforming product and adequate performance margins.

On the other hand, for vendor-supplied off-the-shelf products, quality engineers or experienced quality professionals can perform an adequate evaluation. All surveys should be conducted under the jurisdiction of and with the participation of the procurement function (purchasing, materials, etc.) Providing the supplier with clear requirements, an accurate purchase order defining drawings and applicable specifications, and price and delivery information is essential. As technology advances, computer usage to develop (CAD) designs offers an opportunity to prevent marginal performance, thereby enabling product to be designed in a more defect-free manner. When this situation exists and a vendor is needed, one must ascertain the vendor's equipment can accept the software provided by the computer-aided design (CAD) facility of the procuring company or else that an adequate post-processor program is available to convert data to usable programs for the suppliers' tools and equipment. Standard protocols exist to enable dissimilar computer systems and software to communicate. Initial geometry exchange specification is one such protocol, but you can be assured that other and betters ones will be developed.

It is possible to transmit CAD data via modem and telephone link directly to the supplier. Compatibility of these systems must be verified beforehand to assure proper receipt and conversion of the transmitted data. Error-correcting modems can be used for this purpose. As technology progresses, more capability will be available and quality systems must be modified accordingly. Currently, CAD systems interface with many computer-aided manufacturing (CAM) Systems as well as with computer-aided test (CAT) equipment and computer-aided quality (CAQ) machines.

This technological growth can lead to the production of defect-free product more quickly and at lower cost provided that proper controls are instituted. These entail proper and current configuration controls, statistical process control, purchase order placement, conversion of purchasing documentation to manufacturing orders, availability of compatible software, and so on. One aspect of vendor partnering is the ready exchange of electronic files. Larger companies have developed the necessary software and can demand that suppliers conform. One such system is Manufacturing Automation Protocol (MAP) developed and in use by General Motors. Similar systems are available to smaller companies and must be used to effect tight coupling among design, manufacturing, quality, and production control between the procuring company and the supplier.

In broad terms, the vendor should be evaluated for financial stability, facilities and equipment, quality system, organizational impact of quality, personnel competence in all relevant areas, and manufacturing experience and skill and management. There is a tendency to be overly concerned with detailed procedures while overlooking more important systems aspects. This should be recognized and addressed in the survey. From a quality perspective, some questions to ask include:

1. How will the vendor manufacture your product?
2. What are the points of control?
3. Where are the in-process check points?
4. What are the vendor's process capabilities?
5. What is the vendor's visibility into process control results?
6. What type of equipment is used for manufacturing (computer controlled or not)?
7. What yields is the vendor experiencing on similar products?
8. What methods are used to control test and inspection characteristics of the specification you are imposing?
9. What level of automation is used in manufacture and quality controls (inspection and testing), how effective is the corrective action program, how do they control rework and repair, and how do they control configuration?

A review of the facility including the manufacturing process should provide insight into many items: for example, how much rework is being done. Piles of rework is evidence of poor manufacturing, lack of management awareness, poor process capability, poor attitude, or a combination of these. Insufficient process control may evidence itself in low final yields. On the other hand, processes that perform well within specification limits may not require many, if any, controls. The key lies in how capable the process is in meeting the specification. This is reflected by the ratio of the tolerance (USL-LSL) divided by the natural capability, 6σ, of the process and is designated C_p. A minimum ratio of 1.33 is desired. The results are also reflected in total quality costs, which the vendor may be reluctant to share with customers. In all likelihood, the vendor does not maintain quality cost records. But large accumulations of rework, low final test or inspection yields, and large rework or repair operations can all be indicators of poor performance. Results must be analyzed in the context of other similar manufacturing organizations.

At the same time, there must be an awareness of what is available from other world-class competitors. In the semiconductor business, there was a time when American manufacturers were producing to an acceptable quality level (AQL) of 0.25% while the Japanese were manufacturing to an AQL of 0.005%. although this may no longer be true, it was some time before parity was achieved.

Supplier Communications

Suppliers should be advised where and how the products they are supplying will be used. Aspects of electronic file transfer of purchase order information and requirements should be established early. Reviews of the proposed quality inspection and test points as well as characteristics to be inspected or tested, the frequency of the test, data collection and analysis schemes, and process capability and control should be explored. In some instances, a classification of characteristics into levels of severity is developed. This only serves to create an impression that certain levels of nonconformance are tolerable, and is thus undesirable. It is usually better to determine the acceptance criteria from a functional, measurement, and visual perspective and to identify the characteristics to be evaluated and their limits of acceptability and not to classify characteristics. If there is a prior history of the supplier's performance, this should be reviewed and may, in itself, be sufficient to determine the acceptability of the supplier.

Suppliers should be made to understand that they are an important part of your team. Any means accessible to the procuring activity should be used to accomplish this, including contacts with the vendor's general manager, vice president, or president by an appropriate manager in the procuring company's activity. Although this is normally done by the materials or purchasing manager, it could as well be done by the company's general manager or president. This communication is vitally important to the next step in supplier control and the cost of the procured material.

The results of the survey should be reviewed in depth. Any necessary corrective actions must be identified and scheduled for resolution. All points of inspection, testing, or process control, together with limits of acceptability, must be determined. When product history demonstrates product excellence, no such review may be needed. Each case must stand on its own merits.

Receiving and On-Site Quality Controls

Once an order is placed with a supplier, a decision must be made concerning the degree of control to be exercised on the deliveries. If confidence can be established in the compliance of the item to the specification, product verification can be reduced or eliminated. The ultimate goal is to accept product directly into stores or assembly, with no incoming or field product verification. This goal is achievable but it requires close cooperation between vendor and user.

Until such a situation is reached, some form of product verification must be performed. Often, it is more economical to perform the inspection, test, or control activity in the vendor's facility. Situations making this preferable are:

1. Product quality is largely influenced by in-process operations and controls.
2. Inspection or test cannot be adequately performed on the end product.
3. Purchaser does not have the necessary equipment to test.
4. Product is extremely complex.
5. Purchaser's costs will be less.
6. Evaluation of end product is destructive.

For those situations where incoming product verification is performed upon receipt, the goal must be to eliminate this function. Large companies have incoming quality control departments and make large capital investments in test and inspection equipment to verify parts provided by suppliers because this is the way the business cycle evolved. This situation helps to perpetuate the incoming inspection activity. The entire focus of the organization is on product verification rather than establishing a defect-free product source. Organizational inertia must be overcome to redirect attention to the objective of receiving defect-free product. To a great extent, the Japanese have eliminated this cost by making sure that suppliers' products fully conform to the users' needs.

This situation can be addressed in a systematic way. First, identify the types of nonconformities found in incoming materials. Since the last activity a reputable supplier performs before packing and shipping is final test and inspection, nonconforming product should not arrive at the user's facility. What are the reasons for nonconforming products being received? Some possible ones are:

1. *Acceptable quality level too high.* The promulgation of acceptance sampling using rather high values for acceptable quality levels (AQLs) ranging 0.4 to 2.5% and higher has led to careless manufacturing processes, thus creating lots or batches of products with higher levels of defects than can be tolerated in subsequent manufacturing processes. Processes that meet AQL levels of 1% may contain more than 1% defective and still be accepted (see Chapter 9). For example, if a lot actually contains 1% defective, it will have a high probability of acceptance (on the order of 95%). This results in 10,000 defects for every 1,000,000 parts produced. By contrast, current levels of defects for many parts are being produced at a rate of 50 to 100 defects per million and these levels will improve in time. To overcome this problem, process capabilities more compatible with design must be established. In the interim, AQL values at 0.001% and lower need to be applied.
2. *Out-of-control processes.* Out-of-control processes must be addressed by the vendor's design engineering, manufacturing engineering, and quality engineering groups. There must be an awareness of the problems and a concentrated effort to correct them. Factors contributing to the condition need to be identified and eliminated. Requirements for the vendor to provide customers with process control charts and results of the test and inspection are being increasingly imposed and may be an option to consider.
3. *Inadequate process controls.* Inadequate process controls can sometimes be determined by a good-quality audit. An indicator of poor processes or inadequate controls is consistently high reject rates at in-process or final inspection and test. Another way to determine this is to evaluate test and inspection yields. Some vendors are not aware of these results, and that may be a clue to a design, manufacturing,

inspection, or test deficiency. An effective method for determining process performance is to collect measurement (variable) data and use a histogram (or frequency distribution), control chart, or even a simple graph to display the data in a time series.

4. *Inadequate process capability.* Inadequate process capability must be addressed by the vendor's design, manufacturing, and quality functions. An awareness of this condition can best be determined by statistical quality control procedures. The process capability, C_p, must be greater than 1, with at least 1.3 being desirable. When this situation exists, corrective measures must be ongoing until the process becomes capable of meeting specifications consistently with a $C_{pk} = 1.33$ or greater.

5. *Inadequate final test or inspection.* Inadequate final test or inspection by the supplier could be caused by lack of adequate test equipment, poor calibration, inspection or test error, inadequate specification definition, or irresponsible management. These problems must be resolved. A close working relationship with the supplier is necessary to achieve this. Perhaps field source inspection can remedy the problem initially. Bear in mind that field inspection costs money and should also be a target for elimination. The end objective should be no incoming test, or inspection and no field inspection.

6. *Lack of clear specifications.* Lack of clear specifications can be corrected by properly imposing requirements in advance. It must be recognized that all dimensions on a drawing are rarely checked. Those dimensions termed critical by design engineering (which is rarely done), by quality engineering (which is occasionally done), or by the inspector (which most often is done) are the dimensions inspected. For electrical tests, computerized testing often determines test parameters and limits or specifications list characteristics to be tested. In the latter situation, as the electronics evolution advances, semiconductors devices have become so complex that test equipment costs hundreds of thousands and even a million or more dollars. To accommodate this situation, independent test houses may have to be employed. However, the nature of defects must be analyzed in an attempt to isolate the causes and eliminate them.

7. *High in-process rejects.* The presence of large numbers of rejects at final testing caused by processes not meeting specifications is perhaps the most difficult problem to resolve. In many instances, the process capability may simply be inadequate to meet requirements. The best solution is to utilize statistical process controls and statistical process optimization practices to improve yields. In addition to this, it may

be necessary to use guard bands (test limits tightened beyond specification limits), perform several 100% tests (which is inefficient and should only be done until improvements can be made), apply environmental screens, or use some other technique to remove nonconforming products. Specifying statistical process control requirements is one way to achieve these objectives in a cost-effective manner. Ultimately, they will prevent or reduce the generation of defects.

8. *Poor correlation of test equipment.* Poor correlation of test equipment and measuring for characteristics that have not been specified has been discussed previously. Suffice it to say that calibration of equipment is essential and the calibration system adequacy should be established during pre-award surveys. Care must be exercised to assure that requirements are adequately and completely defined for the suppliers. Users and suppliers should both agree on what will be tested and how it will be evaluated.

9. *Wrong parts.* Wrong parts shipped is simple and could be verified by a quick check of one piece in a lot and prevented by improved vendor quality control. Occasional wrong parts in a shipment is indicative of poor supplier management, housekeeping inadequacy, or carelessness. This must be corrected.

10. *Cheating.* What can be said? Get rid of this supplier just as soon as you can. Litigation is expensive, time consuming, and frustrating but may be your only resort.

Summary

The supplier is an important ingredient in the success of a company. Good suppliers can help in many ways, ranging from improving the design of the item being procured, to specification modifications that enable a process to produce defect-free products, to timely deliveries keeping inventories down and production lines operating. The good vendor has a quality system assuring the user that products will be consistently conforming, and enabling the elimination or major reduction of incoming inspection. Savings realized by this situation can become profits if the receiving quality activity is staffed accordingly. Achieving this objective is not easy, but it must be done in order to survive and prosper in the competitive world marketplace.

References

Bossert, James L., ed. (1988). *Procurement Quality Control,* 4th ed., ASQC Quality Press, Milwaukee, Wis.

Johnson, Ross H., and Richard T. Weber (1985). *Buying Quality,* ASQC Quality Press, Milwaukee, Wis.

Laford, Richard J. (1986). *Ship to Stock,* ASQC Quality Press, Milwaukee, Wis.

Maass, Richard A., and the ASQC Vendor–Vendee Technical Committee (1988). *World Class Quality: An Innovative Prescription for Survival,* American Society for Quality Control, Milwaukee, Wis.

Maass, Richard A., and the ASQC Vendor–Vendee Technical Committee (1986). *For Goodness' Sake, HELP,* American Society for Quality Control, Milwaukee, Wis.

13 · Quality Control Circles

Introduction

One of the exports from Japan is the quality control circle or quality circle, which has been adopted by American companies in an attempt to emulate the success of Japanese product quality and reliability. Started in Japan in the early 1960s, the concept was developed to continue quality improvement and to help overcome the problem of worker lethargy while involving workers in day-to-day problem solving.

Shortly after World War II, Japanese management embraced the concepts of statistical quality control (SQC), currently termed statistical process control (SPC), with the help of American management consultants, most notably W. Edwards Deming and Joseph M. Juran, who began extensive and intensive training in SQC for all levels of management. Quality control literature was broadly available and many training programs were implemented. During this time, American management was casting a wary eye at the quality profession as being somewhat semiprofessional and undoubtedly costing the company unnecessary expenditures. Unfortunately, the quality profession did little to help their image as a business-oriented profession whose objectives were to support corporate goals by producing only conforming product. Whether this was due to the fact that corporate quality functions did not exist, did not have clout, were staffed by managers who were unable to make it in other fields, were too statistically oriented, or were unable to get management's ear for other reasons is unclear. But the fact is that Japanese industry was driven by a desire for superior product quality while American industry waited in the wings. Only after 15 years or so of learning the details of SQC did Japanese management seek to involve their work force in an organized manner to help solve quality problems initially and other types of problems later.

The early quality circles consisted of 10 to 15 workers usually from the same department or work area who put in extra time without pay to work on quality-related problems. Although this is hard to imagine in the United States, the work and social environment in Japan, where workers live in company-supplied living quarter, had job security, and maintained close relationships within the company society, fostered this team spirit. The concept was a natural outgrowth of Japanese culture, where living quarters are generally close, with little room for individual expression and a propensity for conformity and collective thought, harmony, and consensus.

Gradually, the quality circle phased into circumstances that involved meeting on company time and being trained in basic skills of problem solving and statistical quality control. As newer problem solving methods are developed they are introduced into quality circles: the cause-and-effect diagram developed by R. Ishikawa, cause-and-effect diagram and cards developed by Ryuji Fukuda, data flow diagram, KJ diagram, matrix diagram, matrix data analysis, and others. As these techniques, many of which are industrial engineering practices, are developed into working tools of the quality circles, they are incorporated into Japanese circle activities. The definition of the *quality circle* is: A small group of workers doing similar work who voluntarily meet for an hour each week to discuss their quality problems, investigate causes, recommend solutions, and implement corrective action when the authority to do so is within their purview.

To a large extent, the problems that were initially resolved by QC circles were not excessively technical but were important to solve. Later in the cycle of QC circle activity, problems were provided by management or engineering for solution. The fact that unions were largely company unions which, while supporting the workers, recognized the need for harmonious relationships with management and identified the competition and not the management as the "enemy" also helped. The concept worked for the Japanese, worker interest was sparked, and contributions to productivity were many and wide ranging. As the awareness of Japanese management about QC circle benefits grew, applications expanded and the circle movement took off. American management lacked sensitivity as to how circles worked, and unions were determined to avoid give-backs of hard-won work rules and compensation. American management failed to come up with innovative approaches until in many instances it was too late.

A few other factors are worthy of note because of the circumstances that exist in the United States. Japanese companies operate on a bonus system that provides the workers with a direct financial reward based on company productivity. The bonus is a significant part of the financial compensation—on the order of 20 to 25% of annual salary. Also, the engineering community was willing to share its knowledge with the work force. And lower- and middle-level management were not fearful of being upstaged by workers under their control.

Another ingredient in the Japanese equation is the lack of availability of raw materials compared to the resources in the United States. These shortages and the relative cost of raw materials demand that the engineering and manufacturing teams give consideration to conservation and judicious use of materials throughout the design and production cycle. Rejects and scrap are intolerable in the minds of Japanese management. This almost certainly was a driving force lending to more automation and robotics.

So much for history. In its fight for survival, management in the United States recognized the foreign threat to competition too late in some industries to restructure in a way that enables effective competition. But it has become clear that product quality is a strategic weapon in the marketplace. Companies that could buy products at competitive prices that did not require incoming inspection, met the specifications, or had less variability, and consumers who have less frustration with repairs and service calls, were all too willing to buy this superior product quality. Manufacturers began to recognize that improved quality results in improved productivity through less rework, less scrap, less troubleshooting, less retest, less reinspection, lower warranty costs, potentially less inspection and test, and better user satisfaction, leading to increased sales. It appeared that Japanese workers through their diligence and work ethics and use of quality circles, had made this happen. Since the work ethic could not be exported, the focus on quality circles were the specific tools that could be used. Despite the awareness of cultural differences this approach seemed reasonable, and after careful study many companies in the United States adopted QC circles as a management tool. As with every tool, there are right ways and wrong ways of employing them.

Some companies ran afoul of unions that perceived this technique as a way to get more from the workers without adequate compensation. Others perceived it as a threat to their security and control. Still others were wary of this new attitude on the part of management that had not previously displayed much interest in the worker. But in some cases management and unions were able to work together to incorporate some form of quality circles into operations.

Workers may have been suspicious as well. But the focus of attention on worker ideas and problem-solving ability gained some converts. This, when coupled with the voluntary nature of the approach to QC circles, encouraged some workers to participate. The attention paid to the work force was also a stimulus and is reminiscent of the historical Hawthorne experiment. In this Western Electric plant, worker output increased when positive reinforcement was applied, and even when negative reinforcement was applied, simply because workers recognized that they were receiving attention.

But one of the biggest problems was with management. Quality circles demand a participating management style. Many American companies have an authoritative management structure. What the boss says goes—do not bother him or her with petty complaints and footdragging. Middle and lower managers recognized that they would be subject to criticism if a worker or team suggested a good idea in their departments. In addition, the measurements being applied to these managers were schedule, schedule, schedule, and cost. There were no measurements applied to meeting yield requirements and how schedules would be met additional time was spent in training and problem solving. Management on these levels thus saw little reason to accept this new Japanese management tool.

How, then, does a quality circle program get started? It requires a lot of hard work and a genuine change in attitude on the part of upper management, a reorientation of goals, and constant communication among managers. The mechanics of quality circles are fairly straightforward.

The Startup Process

Getting quality circles started requires the use of facilitators whose task is to train the work force in problem-solving techniques and basic statistical quality control and to serve as a link to management to provide the support and resources necessary to incorporate solutions. The process must be carefully explained to the work force and volunteers must be obtained. Each step in the process must be carefully thought out to enable the concept to have a chance of succeeding.

The key to success, however, is the first-line supervision, the foremen, and supervisors who are in day-to-day contact with the workers. The process must be explained to them carefully and their fears allayed. These management team members are concerned over the apparent loss of authority, especially since the process involves presentations by the problem-solving teams to levels of management above them. These supervisory people are also threatened by upper-level managers who question them about why they did not identify problems that their workers developed. It is essential that, by words and actions, the upper-level managers calm the concerns of lower-level managers. The full process must be carefully explained, the rationale developed and presented, and sessions of questions and answers among managers at all levels offered, and lower-level managers must not be criticized for failing to come up with solutions offered by the teams.

Training

Training the work force normally involves managers initially and then the work force. Sometimes workers and managers are trained together. The training itself can range from about 10 to 40 hours. The training can involve the following types of training elements:

1. Brainstorming
2. The cause-and-effect diagram, also known as the Ishikawa or fishbone diagram
3. The Pareto analysis
4. Histograms and frequency distributions
5. Scatter plots
6. Team building
7. Problem-solving strategies
8. Graphs
9. Control charts
10. Cost analysis
11. Flowcharts
12. Other relevant topics, depending on the facility and the type of work being done

Entire books have been written on the quality circle developments and operations. In this section only some aspects will be presented.

Brainstorming

Brainstorming is a process that was developed to evoke new ideas and concepts and allow the pyramiding of ideas. In its elementary form, a group such as a quality circle team states possible causes for the problem under investigation without permitting comment or criticism by other group members. These ideas are listed and as one idea triggers another, these are simply added to the list. In a more structured form each participant is asked to submit an idea. The person is allowed to pass if no idea occurs, and when everyone passes two successive times, the brainstorming is complete. In another scenario, a time limit can be set and all ideas recorded during that time frame are listed. Following this session it is necessary to classify the recommendations into related categories and assign priorities to the most likely causes. In another context, this system can be used to identify the problem that the group wishes to address, rather than a solution to a problem.

The Cause-and-Effect Diagram

One of the more effective tools used by quality circles to classify problem causes is the cause-and-effect diagram developed by Kaoru Ishikawa (see Ishikawa, 1985), also called the fishbone or Ishikawa diagram (see Fig. 13.1). The problem is listed at the head of the diagram, and along the spine are branches identifying the major areas of the problem. Secondary branches then provide the stems, and tertiary items are the twigs. Once this has been formed, the circle group can see the total picture and can identify the most likely cause or causes. This systematic approach helps put together the causes as perceived by circle members.

It is then necessary to select how to obtain data to determine actual performance. Circle groups are usually taught some data-gathering methods and some statistical methods for data analysis to help identify the key contributors to the problem. Sometimes data are already available from the existing quality system. At other times members of the group must go out and collect data. If this is the case, care must be exercised to assure that the proper data are collected and that too many data are not collected. What is too much? Experience will tell, but often a plot of the data can reveal peculiar behavior patterns that are

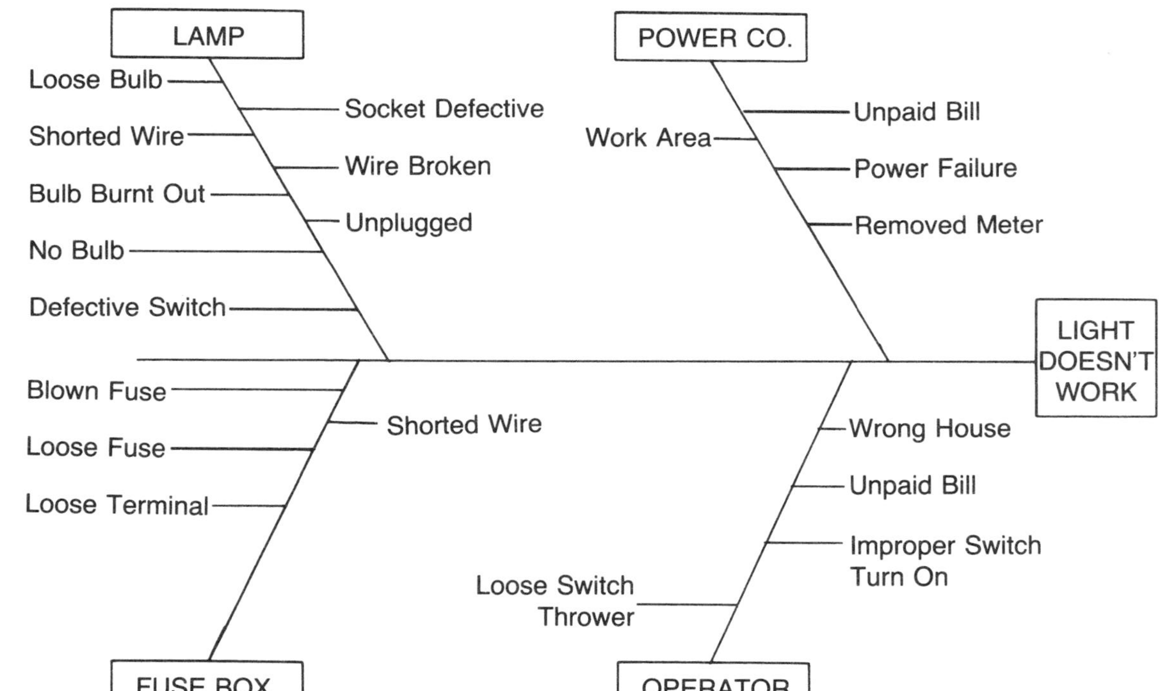

Figure 13.1 Ishikawa diagram (cause-and-effect diagram).

clues to problem causes. When variables measurements are used, perhaps 40 to 50 readings are sufficient, or samples as small as 5 to 10 readings may be acceptable. When attributes data are used, more data are needed. See Ott (1975) and Chapter 6.

The Problem-Solving Cycle

Depending on whose book one reads, there are anywhere from six to nine steps in the problem-solving cycle. Basically, the elements are:

1. Identify a problem.
2. Determine possible causes.
3. Gather data.
4. Analyze the data.
5. Select the most likely solution.
6. Try the solution.
7. If the solution fixes the problem, make sure that it stays fixed.
8. If the solution chosen does not fix the problem, try the next most likely solution; or
9. Ask for help (perhaps a designed experiement is the approach needed).

Problem selection must be controlled to some extent. The Pareto analysis ranks the problems in priority order. It may be desirable to avoid the biggest problem because it may be difficult to solve or needs many resources and may therefore take a great deal of time or never get solved. So a frank examination of the problem is necessary to determine that it is first of all important, as determined by the members, within the scope of the group, and relatively easy to solve. Being within the scope of the group is critical because the role of the circle is to address internal rather than external problems.

Benefits of Circle Activities

The concept of the quality circle—to employ the job knowledge of the work force as well as their intellect to help resolve job-related problems—is ingenious. In Japan it is a natural extension of the participative management style and cultural upbringing. The lifetime employment policies of the larger Japanese firms and their programs of seniority promotions enable the quality circle system to improve worker performance, generate job enthusiasm, and help overcome assembly-line boredom. Quality circle activities have been expanded from mostly quality-related problems, to issues in manufacturing, purchasing, production control, office operations, engineering, and other areas limited only by imagination of the management and the work force. The strategy is to use the human resources to their maximum potential. It should not be forgotten that Japanese management had been exposed to extensive training in statistical quality control for some 15 years prior to the introduction of quality circles.

Differences in American Industry

Significant differences exist between industry and the people in the United States and in Japan. As such it is a tribute to American firms who adapted the concept of quality circles to their management style and made them work. There were, and will continue to be, many less successful situations. An examination of the causes is worthwhile.

American management is, to a large extent, hierarchical in nature. The general need for a participative style of management in operating effective quality circles is anathema in many companies. Management perceives a weakening of their authority and is concerned over the effect of quality circles on this authority. The workers recognize this as well and many feel that this is just another gimmick—which, indeed it is unless the management style changes—and that this, too, will pass in time. These workers feel that if they are just patient, management will tire of this new "quick-fix" scheme. It is no wonder, then, that as Japanese management took over or built firms in America, quality circles were very slow to be introduced.

One of the biggest problems in trying to introduce circles in American industry comes from lower management. These managers, whose day-to-day jobs are in direct contact with the workers, feel most threatened by their perceived loss of authority. First of all, workers are being trained to solve problems within their departments. Why, they reason, doesn't my management feel that I can solve these problems? Is it because I can't do my job? When solutions are provided by workers, will their bosses ask why they have not solved the problem? It is essential that these fears be addressed. Careful and complete training must be provided for these managers and their ideas solicited before quality circles are imposed on them. The diffusion of their authority must be presented in its positive aspects—that it will make their jobs easier. The way in which they are measured must be evaluated. For example, if their performance is principally a schedule-related evaluation, perhaps a quality criterion or problems-solved criterion should be added to the daily or weekly meetings held. There will be an initial loss in output due to training and quality circle activities that occupy the workers' time. This must be recognized up front and tolerated for the longer-term gain that may be realized as more problems get solved.

Japanese executives tend to be generalists with experience in many disciplines, whereas American executives often have experience in finance, law, or marketing and lack an understanding of manufacturing operations. Their perception may be that the Japanese have been so successful because Japanese workers are so industrious, which they are, ignoring the knowledge and participative style of the Japanese executive. Hence quality circles should work there but not here. Although the concept has proven successful in many companies, the old American quick fix does not work; careful planning and deliberation are required to introduce some form of quality circle successfully in this country.

The vast majority of American workers want to do a good job and are very quality oriented. It is the first-line supervisors who set the tone in trying to do what they perceive the boss wants—meeting the schedule. When the supervisor responds to questions concerning product quality by saying something like "Don't bother me about quality, just get the job done—and make it quick," the worker stops raising these issues and works to schedule only.

Actually, the idea behind the quality circles, to use worker knowledge to solve problems, has been used successfully in American industry. The fact that we have a heterogeneous work force can be an advantage over the more homogeneous Japanese work force. Problems can be approached from different backgrounds and a wider variety of solutions can be developed. It is more difficult to arrive at a consensus, but this may not be necessary to solve problems. Given the opportunity, the U.S. worker can innovate solutions. On the other hand, he or she may prefer being the only problem solver rather than sharing ideas with the group. Even though training and organization and development activities show that team problem solving is almost always better than individual problem solving, a feasible mechanism should be provided to allow these workers' suggestions to be provided and to receive prompt evaluation and appropriate rewards.

The mobility of the American work force is another factor that must be considered. On the one hand, it is an advantage because of the variety of training and experience the worker brings to the company. On the other, after receiving training in quality circles or any other discipline, the worker may up and leave before providing sufficient return on the investment. On balance, training is probably advantageous, but this factor must be considered if a decision is to be made on whether or not quality circles are to be used.

It has been the experience of many who have tried that quality circles have a long-term payback, if there is a measurable financial payback at all. It is not a quick-fix solution. It takes months to set up the system, more months to train, and still more months to arrive at and introduce solutions. So if there is a desire to develop the work force into problem-solving units, the limitations must be recognized together with the advantages.

The Reward System

Little has been written about the reward system in the quality circle system. Are the rewards simply those of feeling better because one has made a contribution to the performance of the company? Is a financial reward to be provided if the solution to a problem was developed as part of the quality circle which met on company time and therefore cost the company money to develop? If there is a suggestion system in effect at the time that quality circles were introduced, how the two systems relate must be determined. The decision is not simple and there are no standard answers. But there must be an assessment and a determination made about how this aspect of circle activity is to be handled.

What QC Circles Are Not

The vast amount of publicity received by quality circles may have led management to believe that quality circles are quality programs. They are not! They may be one element in a quality program, not even the major element, but they certainly should not be considered a quality program. Books such as *The Quality System* (Caplan, 1980) more closely define the needs of a quality program.

They are not a solution to union–management problems. There may be some improvement in relations as a result of unions and management working together to develop and implement a quality circle program, but this is only an ingredient in the union–management operation. The circle should not be used to settle grievances or for the resolution of problems between unions and management. Nor should they be used to keep unions out. At best, this is a temporary ploy in trying to resist unionization, and more fundamental approaches are needed to resolve these issues.

The quality circle is not a motivational program. Although its proper use may improve employee involvement, motivation is perceived as much broader than circle activities. Experts say that motivation comes from within the individual and that providing the proper external environment allows these motivated actions to emerge. Many books have been written about motivation; it is my intention only to dispel the belief that circle activity is a motivational program.

Circles cannot address solutions to complex problems. It is possible and desirable to deal with such problems in a task force concept using a multidisciplined team for problem solving. Certainly, statistical methodology is very beneficial in this circumstance, but the statistical methods employed are more advanced than those taught to circle members. Such task forces are not quality circles.

Finally, the quality circle is not an answer to international competition. At best, it is playing catch-up, and if American industry is to emerge as a leader, it must pay leapfrog. The Japanese, and others, continue to improve. As we get better, so do they.

Going beyond quality circles, the concept of the cause-and-effect diagram, for example, has been advanced by Ryuji Fukuda (see Fukuda, 1983), who has added cards to the diagram. These cards, prepared by the workers, express how the procedure or process is carried out in practice and are literally posted to a cause-and-effect diagram to display the current procedures in use. It recognizes that when procedures are written by engineers, they often are not aware of exactly what is taking place. Furthermore, as changes occur, they rarely get introduced into the written procedures. Hence the procedures become outdated and tend to serve as little more than guidelines that are not followed explicitly. The use of cards to update on-floor activities overcomes these disadvantages and highlights process changes, makes the worker the owner of the process, and provides feedback for other processes under development.

Fukuda also advocates a visual control system which provides performance visibility to workers. Placing quality performance charts in areas where workers can see results and compare them to goals or standards is helpful in communicating to operators. Providing information on types and location of nonconformities helps people focus on problem avoidance. The author has had direct experience with this technique and found it to be extremely helpful in improving assembly-line operations.

By studying setup-time problems, Shigeo Shingo has developed a much more rapid machine setup procedure that has enabled more effective short runs to be made, thereby reducing the need to build large inventories to offset the setup time required on certain operations. Dubbed SMED (single-minute exchange of die), the technique uses internal and external factors in machine setup, conversion of internal factors to external factors, and streamlining the internal factors to reduce machine setup time dramatically, from hours to minutes in many cases. Although this may appear to be nothing more than good old-fashioned industrial engineering, it is being put to use to improve productivity in long- and short-run operations.

Summary

Innovations of the type described in this chapter will continue in Japan and elsewhere. The United States must use its resources to create new opportunities for productivity gains. The use of high technology, human resources, innovative management, statistical methods in design and analysis, statistical process control, and other methods must be employed to the level appropriate to gain a competitive edge.

When it comes to group problem solving, management must face up to its responsibilities and establish multidisciplined teams for problem solving, systems managements, and productivity improvements. These team memberships must cut across departmental lines and include people at various levels, from worker to higher manager, to provide an opportunity for success. They must be nurtured by top management to sustain their value.

References

Caplan, Frank (1980). *The Quality System,* Chilton, Radnor, Pa.

Crocker, O., C. Charney, and J. Sik Leung Chiu (1984). *Quality Circles: A Guide to Participation and Productivity,* Facts on File Publications, New York.

Fukuda, Ryuji (1983). *Development of Managerial Engineering,* Productivity Press, Cambridge, Mass.

Ingle, S. (1982). *Quality Circles Master Guide,* Prentice-Hall Englewood Cliffs, N.J.

Ishikawa, Kaoru (1982). *Guide to Quality Control,* Asian Productivity Organization.

Ishikawa, Kaoru, and David Lu, translator (1985). *What Is Total Quality Control? The Japanese Way,* Prentice-Hall, Englewood Cliffs, N.J.

Ott, Ellis R. (1975). *Process Quality Control,* McGraw-Hill, New York.

14 · The Role of the Audit Function

Introduction

As defined in this book, audits are a management tool used to evaluate the quality system. The quality system is not just the quality department, but rather, all operations affecting achievement of the product or service that results in a satisfied customer. When products are audited, the process must entail the selection and evaluation of small samples of product. When services are audited, results of service must be verified with the recipient of the service. When systems are audited, an assessment must be made of the system to appraise the functions being evaluated. Depending on how they are used, audits can make a valuable contribution to operations and to profitability. On the other hand, they can merely be window dressing to satisfy a management directive or some customer requirement. If the latter is the objective, it is advisable to discontinue the audit process and do something more beneficial. The conduct of audits often lacks the necessary ingredients to be effective. The purpose of this chapter is to identify good and poor practices so that more effective audits can be conducted. Financial audits are, of course, essential for effective company operations, but they are not the subject of concern in this book and are not discussed.

Types of Audits

There are basically two purposes of audits. One is to check for compliance to a requirement. Where verification is made, the item or characteristic being audited must conform to the preestablished requirement. The second purpose is to determine the adequacy of a product to satisfy the marketplace, or a service to satisfy a customer. For systems, audits determine whether they are effective and efficient. An example of the compliance audit could be a product performance feature. This is where a sample of product is selected and tested against specified requirements. The sample is representative of the product, and if it meets the requirements, the audit results are acceptable. If it fails to meet the requirements, the audit results are not acceptable and corrective action must be taken. For adequacy, the audit may be an evaluation of the test program. This determines whether the test characteristics or criteria include all the characteristics defined in the specification. If the testing is done in accordance with the test program, the test program is in compliance. If the test program accepts products that do not satisfy the end use, it fails the adequacy criterion.

A further example may deal with housekeeping in a hotel, in which the audit determines whether the housekeeping meets the established standard. The standard, however, must be audited as to whether it is adequate to satisfy the consumer's expectations. A system audit is used to measure the effectiveness of the particular system being evaluated—such as the corrective action system, the purchasing system, or the process control system.

Conduct of Audits

Since it is desired to conduct a meaningful audit, there are a number of factors to consider:

1. What is the area or subject of the audit?
2. Who are to be the audit team members?
3. What checklist is to be used? Is a new or modified checklist in order?
4. How and when will the audit be conducted?

5. What type of report is to be issued? To whom is the audit report issued?
6. What is the audit frequency?
7. What corrective action is required, and when?
8. What follow-up is to be performed?

A discussion of each area, together with items to consider in each, to develop an effective audit process follows.

Audit Subjects

Areas for auditing include subjects for management concern. Frequently, these areas are quality or reliability related, but this limitation should not apply if other functions are a concern. For example, possible subjects are:

1. Control of items procured from vendors and subcontractors
2. In-process and final test, inspection, and quality control
3. Manufacturing documentation and procedures
4. Configuration control
5. Calibration control
6. Reliability program, including failure modes and effects analysis, design reviews, and failure analysis
7. Parts, materials, and process control
8. Program office controls
9. Consumer satisfaction
10. Data collection and reporting
11. Corrective action systems
12. Quality cost program
13. Design adequacy
14. Purchasing
15. Accounting and finance procedures and controls (but not a financial audit)
16. Training program
17. Housekeeping
18. Program office operations
19. Design standards, conformance, and adequacy

This list may serve only as a guide. Those functions important to the company as candidates for audits must be determined on an individual-company basis. When an audit program is established, it sometimes becomes a specialized function in the company. Although there may be advantages in doing this due to skill levels acquired by the auditors, there are some disadvantages as well. For one thing, it becomes an added cost and a function that may tend to perpetuate itself into a situation where the costs outweigh the benefits. Furthermore, the audit itself can become overly complex as the auditors tend to enlarge their responsibilities. It may also be difficult to staff the audit function with talented people who are willing to remain in that capacity for any length of time. It can, however, serve as an effective training ground.

Who Are the Team Members?

When a dedicated function is used to perform audits, there may be a tendency to overlook some aspect of audited functions. While audits tend to look for problem areas, the audit report should provide an objective evaluation of the function being audited. This allows very good areas, products, systems, or services to be identified. As an alternative to a dedicated function audit, team members may be selected from operations to form an audit team. Auditors, whether they are dedicated to the audit function or selected on an ad hoc basis, should have knowledge of the operation being audited, although they do not have to be expert in it. There may even be some advantages to not being an expert because this brings questioning attitude into play. They should also understand the system being audited. When procedures exist, these should be reviewed prior to the audit and should serve as one basis for the audit checklist. The procedure review may help determine the adequacy of the system, although you have to get into the operation to determine how things are really operating. One member of the audit team should be from the operation being audited. He or she could provide the answers to questions and explain how operations are conducted. Since it is natural to "defend one's turf," their responses should be validated independently. Other team members should make sure that there is objective evidence verifying the claims. There may also be a representative of an activity receiving the output of the audited activity. This provides insight into the results obtained from the function. The audit team should be instructed on the conduct of an audit, principally to assess specific samples to evaluate particular requirements; how to use the audit checklist; who to ask questions of, and who not to ask; proper audit reporting; and so on. From two to four auditors may cover any one function, unless the geography is such that more ground must be covered.

Rotation of audit team members has been found to be beneficial. This spreads the task and provides an opportunity for different points of view and different biases to be incorporated into the audit results. When establishing an audit program, serious consideration should be given to a rotating membership on the audit team.

The Checklist

Audits should be conducted using checklists to provide guidelines for the auditors, more so than to limit their audit scope. These checklists should not only be based on requirements in the procedures, but should include elements based on specified requirements. A good source of information for audits are the government quality, reliability, or performance requirements in a variety of Department of Defense, NASA, or Food and Drug Administration standards. Past checklists, checklists from other companies, books on audits and quality systems, or other references can be used as sources for checklists (see, for example, Johnson, 1990). Management concerns that may not be in the standard—often cost-related issues—should be included. For example, in the calibration control area, it is well to determine whether and how often calibration intervals are adjusted and the basis for this adjustment. This particular function sometimes ignores past performance and simply follows prescribed calibration frequency without regard to the needs of the system. It may be that particular equipment rarely requires adjustment during calibration, in which case a lengthened cycle is in order. On the other hand, it may be that the equipment is often out of calibration and the calibration cycle should be reduced or the equipment changed. These data are frequently ignored.

When using previous checklists, it is advisable to alter some of the questions. After several audits have been conducted in an area, the areas where no problems are found should be dropped in favor of new items. The last thing you want to happen is to have a favorable audit when operations are poor. This is indicative of a poor checklist, poor auditing practices, or incompetent auditors. The checklist must reflect meaningful issues. A good rule of thumb is to limit the audit list to about 20 items. Make provision for brief "yes," "no" answers, and provide space for elaboration right on the checklist.

Conduct of the Audit

There are several views on the conduct of an audit. One such view is that the function to be audited should not be forewarned, in order to evaluate it without benefit of a prior corrective action. This may provide a more objective evaluation. The other view holds that the activity or function to be audited should be given advance notice to give them an opportunity to sharpen the conformance aspect of the audit. Some would even provide the audit checklist to enable prior self-audits to be performed. The latter situation is like a take-home exam, whereas the former is like a surprise quiz. Each serves a purpose, and a combination of the two approaches is probably the best way to deal with this consideration. Each approach offers a different perspective, and depending on the situation, each method is valid. The objective is improvement and there is more than one way to get it.

The area to be audited must continue to function during the audit, so it behooves the audit team to review operations with minimum disturbance to the operating personnel. Questions and clarifications should be addressed to supervision, although direct questions to operating personnel may be asked occasionally. During the conduct of the audit, the auditors should be aware of what is happening in an area. For example, in a manufacturing area, shelves provided for rework and the performance of extensive rework operations are indicators that there are problems. If chemicals used during operations such as soldering are unidentified or are obviously old, carelessness and the possibility of other problems being present are indicated.

During the audit, the checklist should be completed and any notations should be made at the time the observation is made. Each auditor should keep his or her own record. Upon completion of the audit, each auditor's comments should be considered. Audit elements should be ranked in order to focus attention on aspects to be considered. Generally, a three-level classification such as significant, major, and minor, or simply one, two, and three, is satisfactory.

The Audit Report

The audit report should be a composite of all auditor's observations. When there is any disagreement among auditors, it is desirable to reach a consensus. By the same token, if one auditor has observed an element that in his or her judgment is neither adequate nor compliant, that comment should be included. The report should have a summary of results presented at the beginning, with audit discrepancies noted in order of significance. The report should be issued to the manager of the activity, with a copy to the area supervisor. It should request corrective action for all discrepant items, but for items that are not safety related, the schedule of corrections should be provided by the audited activity. The major purpose of the audit is to improve operations functionally or economically, so the audit response is crucial and the follow-up action is even more so. If possible, ranking of the activity should be provided to enable auditors and auditees to know the relative assessment so that past and present performance can be compared.

One problem with audits is self-perpetuation, having poor cost/benefit ratios. Once operations are performing as desired, or if problems are of a minor nature, the efforts to enhance operations through audits should be stopped or the frequency of audits should be drastically reduced. Operational improvements can be achieved through other means, such as technology improvements, quality cost analysis, quality improvement programs, quality circle recommendations, or some other approach.

Audit Frequency

The timing and repetition of an audit should be based on the importance of the subject matter to management. Routine repetition should be avoided, since the operation becomes a cost drain and a noncontributor to improved operations. Semiannual and annual reviews should be adequate except where product audits are used to assess product quality. In that instance, monthly audits are usual. Greater frequency than that is indicative of poor management communications to operating personnel. Proper communications by management should result in conformance to requirements without the need for frequent audits.

Corrective Action

The audit report should reflect objective results and praise the good portions as well as to point out poor areas. If the auditors are involved in ongoing audits, they are in a good position to help promulgate clever operating practices to other areas of the company as well as to assess generic problems among various functions in the company.

The key ingredient in an audit program is the corrective action that results. If there is no corrective action or bona fide efforts toward corrective action, the audit program is ineffective. The audit report should point corrective action efforts in the proper direction. If audits are to be meaningful, proper classification of the problems disclosed must be made. The audited activity should provide the auditors with a schedule of corrective action and the general approach planned except for issues of safety requiring immediate administration. When corrections are completed, auditors should be advised, preferably in writing, but some notification of type and extent of corrective measures must be provided.

Follow-up

Follow-up reviews or meetings must be held to determine the adequacy of the corrective action. When completed satisfactorily, the audit team should then determine an approximate reaudit time. The frequency should be adjusted based on severity of problems disclosed and the significance of the function or system being audited.

There are times when corrective actions provide elegant solutions to problems. The auditors may be able to benefit the company by spreading to other areas the word on good corrective measures. Since the costs have already been incurred in the conduct of the audit, the benefits might as well be maximized.

Field Operations

One valuable source of information is field operations or field service activities. These operations can provide valuable information on problems in the field so that design, manufacturing, or some other function can take action to eliminate the problem. Unfortunately, field operations are usually divorced from the quality function and the field operations manager regards his or her primary task as getting the product fixed—not as providing data to the plant. A world-class operation thrives on information. Field information is the culmination of the entire effort and the data must be used for continuous self-improvement.

Service Quality

A special mention of service quality is worth noting. Product audits can evaluate products in the factory, warehouse, or in the field, although services can be evaluated only after they have been performed, and the quality of service is in the view of the person serviced. In this case, telephone or mail is the fastest and lowest cost but not necessarily the best way to gather information. Think about how many times you have responded to mail or questionnaires, or have hung up on telephone callers. In some situations, customer service operations can develop assessments of performance, although it may be biased. After all, customer service departments rarely get calls or letters complimenting them on a service. A more effective way of data gathering uses selected sites to monitor results. These must be developed, nurtured, and used to effect improvements.

Summary

An audit program is another tool to improve operations, but its effectiveness is subject to management desires and controls. As in most operations, care should be exercised to avoid "going through the motions" to satisfy either a customer or management. Be honest with yourselves. Make the effort pay off. Avoid routine, examine the nature of deficiencies disclosed, and take action to preclude recurrence. Use the audit results to adjust audit frequency or eliminate specific audits, adopt innovative solutions to problems, and share generally good methods with other segments of the company. In short, if you perform audits, make them pay off or stop performing them.

References

ANSI/ASQC Q 90–1987, *Quality Management and Quality Assurance Standards: Guidelines for Selection and Use,* American Society for Quality Control, Milwaukee, Wis.

ANSI/ASQC Q 91–1987, *Quality Systems: Model for Quality Assurance in Design/Development, Production, Installation, and Servicing,* American Society for Quality Control, Milwaukee, Wis.

ANSI/ASQC Q 92–1987, *Quality Systems: Model for Quality Assurance in Production and Installation,* American Society for Quality Control, Milwaukee, Wis.

ANSI/ASQC Q 93–1987, *Quality Systems: Model for Quality Assurance in Final Inspection and Test,* American Society for Quality Control, Milwaukee, Wis.

ANSI/ASQC Q 94–1987, *Quality Management and Quality System Elements,* American Society for Quality Control, Milwaukee, Wis.

Johnson, L. M. (1990). *Quality Assurance Evaluator's Handbook,* Stockton Trade Press, Stockton, CA.

15 · Reliability in New Product Planning

Introduction

No discussion of world-class quality would be complete without emphasizing product reliability. *Product reliability* is the probability that the product will perform its intended mission for a specified or predetermined time in the intended environment. For the consumer, this translates into a trouble-free product that requires little or no maintenance during its normal service life. While it is expected that a high-quality product will perform dependably over a long period of time, the achievement of highly reliable products requires somewhat different approaches than does the production of a defect-free or high-quality product as it leaves the manufacturer.

Providers of goods and services have already seen longer-term product warranties used as a marketing tool in electronics and automobiles. This trend is bound to continue. Indeed, those products that will last as long as the consumer is satisfied with them, providing that they are not abused and are properly maintained, are the products that will dominate the marketplace. The mark of a professionally executed product is its reliability. Although the Japanese automobile captured a large share of the American market in the early 1970s because they were fuel efficient during a period of skyrocketing fuel prices, they retained and improved their market share because the product was very reliable—more reliable than their American counterpart.

Consumer demands and competition requires that reliability be given a major role in product development. Product reliability must be considered throughout all phases of the product cycle, from design through field operation. What needs to be done in each phase of the product cycle is described below.

Design Phase

During the design phase, reliability requirements, parts, materials, and process standards must be determined, derating practices must be established, and design review and product safety requirements must be implemented. Each product should have reliability goal established in terms of *mean time between failures* (MTBF), which is based on the desired reliability, cost per service call, or other market considerations. A determination must be made concerning the requirements for initial performance and the average number of service calls during 1-, 2-, or 5-year periods, as well as the projected design life of the equipment. Multiple-year analyses are required because consumer expectations extend to that realm. Furthermore, competitive advances may accrue should extended warranties be economically achievable. TV sets may be warranted for 1 to 3 years but many last for 10 years. Appliances may be warranted for 3 to 5 years but many last for 10, 20 and even more years. But manufacturers are reluctant to warrant products for that period of time because of the potential cost impacts. In recent years, automobiles have had warranties extended from 3 years to 5 to 7 years as a sales strategy, and it is likely that other products will follow suit. The manufacturer that figures out how to do this economically will force others to conform. Product weaknesses must be known to enable an intelligent design decision to be reached that provides an economical extended warranty.

To achieve equipment reliability, equipment should be partitioned to respective design areas to achieve an allocated and specified MTBF. This must be a design criterion just as any other performance criterion specified for the product. It is attained through careful part, material, and process selection, coupled with mathematical analyses of individual part contributions to the overall reliability requirement. The means to assess the achievement of specified reliability, goals must also be planned at this early stage of product life.

To be reliable, a product should use parts, materials, and processes based on past history, engineering knowledge, and manufacturing experience. Process capabilities of the manufacturing organization must also be considered. New processes should be proven or qualified before use. Too many times unproven processes have created costly production and yield problems. Parts should be selected from approved standard parts lists, materials should be selected from standard materials lists and known processes should be identified. Controls should be established of the application of nonstandard parts, materials, and processes so that adequate checks are made to make sure that early life failure mechanisms are not inadvertently built into equipment. New parts, materials, or processes should be thoroughly evaluated before incorporating them into new products. This is particularly important because the use of custom large-scale integrated circuits (CLSIs) or application-specific integrated circuits (ASICs) electrical designs can offer major reliability gains if the design elements and processes used in their manufacture are themselves reliable.

An equipment derating policy should be established for each piece of equipment and provided to design groups so that the derating policy will be followed. Where design analysis indicates thorough criticality, analysis should be done or measurements should be taken during design to increase the likelihood of arriving at a design that achieves the specified reliability goal. Techniques such as redundancy should be employed, when necessary, to achieve the specified reliability.

Other techniques, such as failure mode and effects analyses, stress testing, use of Duane plots, and more, defined in various reliability texts, will help achieve reliable product performance. All products should be subjected to design reviews. These should be conducted as a preliminary design review to ascertain the readiness of the product to proceed with design and a critical design review which occurs prior to entering production.

All action items should be resolved from each review prior to moving to the next phase. Review participants should include design experts who have not actually participated in the design, as well as experts in reliability, maintainability, quality, production, marketing, and field services. The thrust of the review should be to question design integrity to assure that reliability, quality, producibility, and maintainability are designed into the product. The use of computers to provide functional simulation can be very helpful in reducing design risks. The use of statistical experimental designs to evaluate alternative designs and establish "robust" product design is highly desirable. *Robustness* is that property that results in stable performance over a wide range of environments and to the end of life. Products with these characteristics enable cost-competitive production with high quality and good reliability and are essential to compete in the world marketplace. Computer programs including expert systems or even artificial intelligence are beginning to be available.

Safety considerations should be taken into account during the design and reviewed to assure that the product will function normally in a manner that will deliver expected service with no unreasonable hazard to the user. Anticipated misuse during operation should be considered to assure that product safety is maintained throughout design, manufacture, and field use. The design should incorporate features that enable proper maintenance and service to be performed during field operations.

Product design is the major contributor to developing world-class products. Products other than electronic products may be dealt with differently. There are other considerations that must be taken into account. Shelf life and stability are examples of such characteristics. A means for verifying these product attributes must be considered. Whatever product is being developed, performance throughout its operational period should be considered and means for dealing with "reliability" should be addressed.

Preproduction Phase

Prior to release of product to the factory, a qualification testing specification should be developed which, as a minimum, identifies extremes of environment to which the equipment will be exposed. For electronic equipment, a test program utilizing temperature or temperature cycling in combination with vibration testing should be conducted to provide data for verification of the reliability specified. Test levels should exceed use environments by some margin. Margins should be great enough to stress the part substantially more than the use environment but not so great that unusual failure modes are created that are unlikely to occur in application. When the design group performs these environmental tests there is a great temptation to make adjustments as testing proceeds—thus masking potential problems. Because equipment should be operated *without* readjustment of factory-adjustable elements, it is best to use an independent group such as the quality function to perform these tests.

All noncompliances should be identified, analyzed, and corrected. A decision should be made as to whether the corrections are necessary in the design or manufacturing cycle. In addition to engineering, design assurance (reliability), quality assurance, and marketing should participate in this decision. Ideally, production should not begin until all corrections have been incorporated and the specified reliability has been achieved. It is particularly important to perform a thorough analysis of all failures, since underlying design weaknesses or part, material, or process limitations may be disclosed during this time.

This is a complex process since sample sizes used to evaluate design adequacy are limited by time, facilities, and cost. But this aspect of design qualification is important and can provide assurances about product integrity. In instances where schedule constraints require production prior to establishing confidence in the ability of the equipment to achieve the specified reliability, careful records must be maintained on early production and required Engineering changes should be incorporated in those products as dictated by the qualification test program. As necessary, preconditioning (or burn-in) may be incorporated for parts or assemblies, or thermal cycling for electrical boards and systems should be considered to achieve the specified reliability. If it becomes evident that the required reliability will not be achieved, a management decision is required to define the course of action.

Procurement Phase

Requirements for warranty consistent with total product warranty should be imposed on suppliers of assemblies, regardless of whether equipment is off the shelf or vendor designed. In the case of off-the-shelf equipment, parts lists should be analyzed to preclude incorporation of parts with marginal life or environmental performance. Other elements of the reliability program should either be imposed on the supplier or performed to assure that the system will achieve its specified reliability.

Subcontractors should be evaluated to assure capability of meeting these requirements. Quality assurance should perform this evaluation as a member of a team or, perhaps, individually, depending on the particular circumstance. Suppliers that are disapproved should require a special review to determine an appropriate course of action. These may include finding another supplier, requiring a redesign, providing support to the supplier during the design or manufacturing phases, or agreement by the supplier to validate product performance.

By evaluating potential suppliers or subcontractors far enough in advance of needs, risks can be minimized. Using advanced technology, such as a CAD/CAM system for manufacturing by suppliers, some risks may be reduced. Advanced technology can be a major factor in design, manufacture, and product verification.

Production Phase

A quality plan should be developed for each product, which, as a minimum, identifies the points of inspection and test, the acceptance criteria, the type and method of data collection and analysis, and a corrective action system. These aspects of manufacturing have been discussed elsewhere in this book. Keep in mind that this plan is dynamic and changes to accommodate differences in results. The thrust should always be toward narrower process capabilities and less inspection and testing.

One particularly important aspect of the reliability program is that of failure analysis. These analyses may indicate potential failures due to physical limitations of the product or improper part, material, or process applications. These analyses may indicate potential failures due to physical limitations of the product or improper part, material, or process applications. This analysis is usually impractical to carry out for each nonconformance that occurs during production. Instead, the dominant recurring problems must be analyzed promptly, while those that occur occasionally or rarely may not be analyzed in as timely a fashion, due to economic and time factors. However, as the underlying causes for the most frequently occurring problems are found and eliminated, those problems that do occur less frequently become the dominant factors and are then subject to detailed analysis and correction. The decision on what gets analyzed is crucial and is also influenced by the consequences of the anomaly. That is, even if an event occurs rarely but its effect is serious, analysis and correction action must be taken promptly. Ideally, all anomalies should be analyzed immediately. Unfortunately, this is usually not possible, due to resource limitations. However, proper design and preproduction activities can keep these nonconformities to a low-enough level to enable each event to be addressed promptly.

Personal experiences over many years of working in and observing other factories leads me to the conclusion that processes are rarely optimized, even after long periods of use. Depending on the type of process, some form of process experimentation is essential to continue to improve yields and reliability. Designed experiments, response surface methods, and evolutionary operation are some of the tools that can be used to improve processes (see Chapter 10).

Operational Phase (Field Use)

A product support plan should be developed for new products or new markets. This plan should identify, as a minimum, who will perform field maintenance and repair, what the spares complements should be, what the spare levels should be, what type of support documentation is necessary, where repair locations should be established, what field performance date will be collected, and to whom the data are to be sent for analysis. One of the major shortcomings of field support operations is that the operation is considered to be primarily to keep the customer satisfied by repairing the deficiency promptly. The information gained from knowing the nature of the field problem is just not used to correct design and manufacturing problems. These data can be invaluable in providing insight to enable improvements in design, manufacturing, or elsewhere in the product cycle to prevent recurrences. This aspect of field operations is vital to successful world-class competition and must be performed thoroughly. One possibility is that field support operations report to quality personnel to force data feedback.

Service Reliability

Service reliability is not quite as well developed as product reliability. How good and how long-lasting is the service being performed? As in product reliability, customer satisfaction with the service and its permanence is a measure of service reliability. If a copier machine requires weekly lubrication to keep it operating efficiently, that may be indicative of poor service or poor design. The service group is criticized, but the selection of the bearing or the lubrication methods, or the lubricant itself, may be the problem.

In responding to a customer inquiry, the speed and the permanence of the problem resolution may be the dominant factor. Each area has its peculiar characteristics and must be evaluated individually. Frequently, the systems and procedures that are established will facilitate service reliability. Results are more difficult to summarize because information must be obtained from those serviced, while for product reliability, more tangible results can be gotten from in-house testing, field performance results, warranty costs, or a combination of these.

People are the key, as they are in most operations, but user (employee)-friendly systems can make results more timely and useful. Here, too, technology plays an important role. Computers, telecommunication of data, and data summaries can all be useful. The key ingredient is management's recognition that something needs to be done to assure reliability in services.

Summary

The suggested program in this chapter is intended to provide a means of achieving minimal warranty expenditures while meeting customer's expectations. The overall cost of the program should be substantially below the warranty expenditures that would otherwise be experienced. Designing and manufacturing a reliable product is essential to competing on a global basis. Those companies that learn how to produce a long-lasting, trouble-free product at reasonable cost are those that will survive.

16 • Epilog: Where Do We Go from Here?

If you are reading this chapter, chances are that you have read something in the text that has struck a responsive chord and you are interested in what to do next. My strongest advice is: "Do something!" We are not making it the way we have been going, and something must be done. The worst advice that I have heard is, "If it ain't broke, don't fix it." Let me assure you: If it ain't broke, it most certainly is not optimized. Unless we continue to work toward improvement, somebody else is out there figuring out ways to produce the products or perform the services better, cheaper, or faster. That somebody will succeed unless you keep working to improve the products and services that you are providing and to maintain your lead or gain on the current leaders.

Management's job is not just to produce a product or a service. They must also produce information that can be used to learn about the process so that it can be made more efficient—thus turning out more with less. Figuring out how to collect these data easily and simply and present it in a form that conveys intelligence is the first challenge. Figuring out how to organize to use the data then becomes the second challenge. Finally, figuring out the structure and communications links to effect smooth, rapid, integrated change is next.

There is no doubt that quality (of product or service) is a key in achieving world-class competitiveness. Designs and manufacturing processes must be compatible. Specifying tolerances that cannot be met by existing equipment forces costs up and quality down. If tighter tolerances are indeed necesasary to meet competition or satisfy a customer, and existing processes cannot deliver, it is necessary either to develop processes that can meet requirements or to accept the added cost of inspection and rejects until ways can be found to bring the tolerance requirements and the process capability into agreement. Using statistical methods to accomplish this, through statistical process control, process improvement, or statistically designed experimentation points, the way to change the processing parameters is a very cost-effective way to proceed.

If you are in a service industry, the procedures used to direct the service to the point of need, the efficient accomplishment of that service and its satisfaction to the customers are essential features. Keep in mind that many services, particularly those that repair something, have more requirements than just satisfying the customer by getting equipment back into operation. There is a wealth of information that can be used to improve the way in which the service is offered or in which the product performs. Providing information to the designers, manufacturing, software developers, or business system analysts can help refocus the company. The challenge is to design the problems out of the system so that there is not a constant need for the type of service being performed, or at least, the need is shrinking.

Remember, too, that once the process of change is started, it must be integrated into the way you do business. Where you start will depend on where you are in your organizational development. How you start will depend on the type of management in place. The authoritarian will start in one way—and perhaps demand participation and begin to measure actions to satisfy his or her demands. The participative manager will organize teams to get some action started. Perhaps there will be an executive council to set policy and project objectives. Below this council there may be a management group, and below them, working teams where the real work gets done. Or maybe some other approach will be developed. Although I favor the participative style, the authoritarian approach can also work provided that there is acceptance of the recommendations for improvement and no retribution for mistakes.

The overall strategy must be to use the knowledge base residing in the personnel in the organization, to provide an outlet for the creativity that exists, and to cultivate the creativity. Make it easy for change to be proposed and nurture the creative talents in your organization by seeing to it that change is allowed to take place. Deming calls this "constancy of purpose" and says that it encourages "never-ending improvement." He also counsels "driving out fear" by allowing differing opinions to be aired—in fact, encouraging them. This is more difficult for an authoritarian management to encourage this behavior than a participative management, but it is not impossible. So let your people help raise your organizational efficiency.

The most significant changes affect more than one functional area. Thus multidisciplined teams tend to be more productive than single functional teams. When organizing for improvement, these multidisciplined teams can address cross-functional problems and resolve them. Single functional teams can address those problems within their purview. But both have a place in the improvement process—and both should be used.

Let us close by looking at some specific ideas on how and where to get started.

1. *Recognize opportunities to do a better job in all business areas.* Your operation may not be broken, but there is most certainly a better way, especially if you consider the new technology available to get your job done.

2. *There are existing points of natural data collection.* Use test and inspection data to improve the processes that are used to create the product that they are testing or inspecting—not just to fix the nonconformances. Use field data to improve your designs, manufacturing, software development, or business systems. These data must be accurate, timely, and in a simple format that provides visibility into what is really happening. Later, you may need other data-gathering points, but start with what is already in place.

3. *Organize for quality improvement (or corrective action).* All these data must be put to use or else, stop collecting them. So individuals or teams must be set up to meet regularly and provide a forum for action: whether a 10-minute stand-up meeting every morning or a once-a-week review. But this process must be a part of any continuous improvement activity. There should also be a focal point for quality and productivity improvement. The current trend is for a director of quality and productivity, and this may be logical for your organization. A lot depends on the person selected for the position. He or she must be perceived as a contributor. It cannot be someone who has not made it elsewhere and has been placed in this position to finish out a career.

The troops know what is happening. While you're at it, make sure that the title is on a par with other executives at the same level. Because if you use the "director" and everyone else is a vice president, you are sending a message to the people: low-level title, low-level emphasis; the boss does not really mean it—just wait a while and a new fad will emerge.

4. *Correct the problem by fixing the process.* Simply fixing the product is not corrective action or quality improvement. It merely gets the products to conform, but unless the process is fixed, more nonconforming products will be generated. You should fix the process even when it does not appear to need fixing (but do not give top priority to that activity). Do not wait for a problem to develop before taking action. If defect levels are too high, use some of the quality improvement methods suggested in this book to reduce them. If a process is in control, make sure that the control is at the right level.

5. *Establish goals to reduce the process variability or increase the yield.* You must have goals to provide a yardstick against which measurements can be made. There are schedule goals and cost goals. Unless you establish quality goals, the other goals will take priority and quality (yield) will remain an afterthought. Measure people's performance against quality goals as often as you measure performance against schedule goals. This orientation must be achieved, and in most organizations, it is a change in thinking for your managers. This change can be accomplished only with ongoing and frequent performance appraisals against quality goals.

6. *Plan for continued improvement.* Establish long-term goals that enable the company to have a vision of where they are headed. These long-term goals must include specific objectives for quality and productivity and they must be measurable, quantifiable, and have a direct bearing on business performance. There appears to be a beginning, in this country, for some companies to set these long-term plans in place. Every business must have them, and product or service quality is a key factor for competitive effectiveness in the world marketplace.

7. *Reexamine the way things are.* There need not be perceived problems to make improvements. But the explosion of technology is bound to affect your business. Keep tabs on the opportunities to do this with more integration of activities, such as the use of common data bases for a whole host of computerized activities—engineering, design, manufacture, test, production planning, inspection, and more. These offer a great opportunity for increasing efficiency. Reducing the time from concept to delivery with high quality can get the job done at

lower cost. All business cycles must be shortened. Whether it is the time required to generate a purchase order and receive material, the time required to process a sales order, or the time required to introduce a new product or service, the time to consummate the cycle must get shorter and shorter with greater and greater accuracy.

8. *As improvements are introduced, adjust the system.* Two simple examples illustrate this. Reject rates were improved in a hearing-aid manufacturing plant, but the rework department still maintained its six-person staff; they just took longer to repair fewer items. The improvements that were made did not have the economic impact that they could have had if four of the rework operators had been put back into production operations. In a semiconductor operation, yields were improved dramatically due to a process change. The five visual inspectors were reduced to two inspectors, and the three other inspectors were converted to production operators making product. A double benefit resulted: products improved and more output resulted not only from yield improvement but also because there were more operators making products. The organization must be dynamic and change as improvements are made.

9. *Ask the right questions and continue to ask them.* Constant vigilance is necessary to institutionalize the process of continuous improvement. It must become ingrained in the way you do business or your company will lose market share to those who are more persistent and continue to hold the leadership torch of quality.

10. *Finally, train, train, train your people in problem solving, dealing with change, and reacting as individuals and teams.* Teach how statistical methods can be used to run your business. This is an information society. Gathering that information in a way that provides communications to the people who can take action and making them aware of how to perceive data so that they lead to intelligent action requires training. So on to the task—and go get 'em!

Epilog/Postscript

The Malcolm Baldrige National Quality Award

This award, named for the late Secretary of Commerce, was established as a part of the Malcolm Baldrige National Quality Improvement Act of 1987, Public Law 100–107, signed by President Reagan on August 20, 1987. It provides for two awards for each of three different types of businesses:

1. Large manufacturing companies
2. Small manufacturing companies (fewer than 500 employees)
3. Service industry

The criteria for these awards are updated annually and contain elements that reflect the current thinking about quality as a strategic issue in business. Copies of the criteria and rules for application may be obtained from

The Department of Commerce
National Institute for Standards and Technology
Gaithersburg, MD 20899

There are many issues addressed in the application. They are subdivided into seven categories:

1. Leadership
2. Information and analysis
3. Strategic quality planning
4. Human resource utilization
5. Quality assurance of products and services
6. Quality results
7. Customer satisfaction

An excellent way to establish areas in which improvement can be accomplished can be developed using these criteria. The contents of this book can help with some methodology and by providing insights as to how effective change can be introduced.

Index